Alternative Solvents for Green Chemistry

RSC Green Chemistry Book Series

Series Editors: James H Clark, *Department of Chemistry, University of York, York, UK*
George A Kraus, *Department of Chemistry, Iowa State University, Iowa, USA*

Green Chemistry is one of the most important and rapidly growing concepts in modern chemistry. Through national awards and funding programmes, national and international courses, networks and conferences, and a dedicated journal, Green Chemistry is now widely recognised as being important in all of the chemical sciences and technologies, and in industry as well as in education and research. The RSC Green Chemistry book series is a timely and unique venture aimed at providing high level research books at the cutting edge of Green Chemistry.

Titles in the Series:

Alternative Solvents for Green Chemistry
By Francesca M. Kerton, *Department of Chemistry, Memorial University of Newfoundland, St. John's, NL, Canada*

The Future of Glycerol: New Uses of a Versatile Raw Material
By Mario Pagliaro, *CNR, Instiute of Nanostructured Materials and Institute for Scientific Methodology, Palermo, Italy and Michele Rossi, Department of Inorganic Chemistry, University of Milan, Milan, Italy*

Visit our website on www.rsc.org/books

For further information please contact:
Sales and Customer Care, Royal Society of Chemistry, Thomas Graham House, Science Park, Milton Road, Cambridge, CB4 0WF, UK
Telephone: +44 (0)1223 432360, Fax: +44 (0)1223 426017, Email: sales@rsc.org

Alternative Solvents for Green Chemistry

Francesca M. Kerton
Department of Chemistry, Memorial University of Newfoundland, St. John's, NL, Canada

RSCPublishing

ISBN: 978-0-85404-163-3

A catalogue record for this book is available from the British Library

Published by The Royal Society of Chemistry,
Thomas Graham House, Science Park, Milton Road,
Cambridge CB4 0WF, UK

Registered Charity Number 207890

For further information see our web site at www.rsc.org

Preface

Everyone is becoming more environmentally conscious and therefore, chemical processes are being developed with their environmental burden in mind. Of course, this also means that more traditional chemical methods are being replaced with new innovations. This includes new solvents.

Solvents are everywhere, but should they be? They are used in most areas including synthetic chemistry, analytical chemistry, pharmaceutical production and processing, the food and flavour industry and the materials and coatings sectors. But, the principles of green chemistry guide us to use less of them, or to use safer, more environmentally friendly solvents if they are essential. Therefore, we should always ask ourselves, do we really need a solvent? Chapter 2 explains some of the challenges and successes in the field of solvent-free chemistry, and the answer becomes apparent: not always!

In the introductory chapter, some of the hazards of conventional solvents (e.g. toxicity and flammability) and their significant contribution to waste streams are highlighted. The general properties of solvents and why and where they are used are outlined. Additionally, EHS (Environmental, Health and Safety) assessments and life cycle analyses for traditional and alternative solvents are described. It becomes clear that often a less hazardous VOC is available and that although only "light green" (or at least "less black") in colour, they can be used as an interim measure until a more satisfying option becomes available. In each of the subsequent chapters, where possible, the use of an alternative solvent is described for a range of chemical applications including extractions, synthetic and materials chemistry. At the beginning of each of these chapters, some of the advantages and disadvantages of that medium are laid out.

Water is often described as Nature's solvent; therefore Chapter 3 describes the solvent properties of water. It is already used quite widely on an industrial scale, particularly in emulsion polymerization processes and hydrodistillations. However, some of the most exciting results have come in the field of synthetic

RSC Green Chemistry Book Series
Alternative Solvents for Green Chemistry
By Francesca M. Kerton
© Francesca M. Kerton 2009
Published by the Royal Society of Chemistry, www.rsc.org

chemistry. Recently, 'on-water' reactions have shown that hydrophobic (water insoluble) compounds can achieve higher rates dispersed in water compared to reactions in conventional solvents or under solvent-free conditions. Water can also be used at very high temperatures and under pressure in a near-critical or supercritical state. Under these conditions, its properties are significantly altered and unusual chemistry can result. This is further discussed in Chapter 4, which describes supercritical fluids. The focus here is on the non-flammable options, that is, carbon dioxide and water. Modifications that are performed on substrates in order to make them soluble in supercritical carbon dioxide are described. Additionally, the benefits of the poor solvating power of carbon dioxide, e.g. selective extractions, are highlighted and its use in tuning reactivity through its variable density is described.

In addition to water and carbon dioxide, there is an increasing availability of solvents sourced from renewable feedstocks including ethanol, ethyl lactate and 2-methyl-tetrahydrofuran. The properties of these solvents and their potential as replacements to petroleum-sourced solvents are discussed in Chapter 5. Renewable feedstocks and their transformations are a growing area of green chemistry and they have significantly impacted the solvent choice arena. In addition to renewable VOC solvents, non-volatile ionic liquid and eutectic mixture solvents have been prepared from renewable feedstocks and are looking to be very promising alternatives in terms of toxicity and degradation. These and other room temperature ionic liquids (RTILs) will be discussed in Chapter 6. The field of RTILs has grown dramatically in the last ten years and the range of anions/cations that can be used to make these non-volatile solvents is continually expanding. Although some of these media may be more expensive than other alternatives, the chance to make task-specific solvents for particular processes is very exciting. RTILs, alongside fluorous solvents, have also made a large impact in the area of recyclable homogeneous catalysts. Fluorous solvents, as described in Chapter 7, show interesting phase behaviour and allow the benefits of a heterogeneous and homogeneous system to be employed by adjusting an external variable such as temperature. Recent advances in this area will be discussed, for example, supported fluorous chemistry, which avoids the use of large amounts of fluorous solvents and might be more amenable to industrial scale processes.

Possibly the least explored and newest options available to the green chemist are liquid polymer solvents (Chapter 8) and switchable and tunable solvents (Chapter 9). Unreactive low molecular weight polymers or those with low glass transition temperatures can be used as non-volatile solvents. In particular, poly(ethyleneglycols) and poly(propyleneglycols) have been used recently in a range of applications. Probably the most important recent additions to our toolbox are switchable solvents. New molecular solvents have been discovered that can be switched from non-volatile to volatile or between polar and non-polar environments by the application of an external stimulus. Gas-expanded liquids will also be discussed in Chapter 9, as carbon dioxide can be used as a solubility switch and to reduce the environmental burden of conventional solvents.

Although many advances in the area of alternative solvents have originated in academia, many alternatives are already in use on an industrial scale. For example, supercritical carbon dioxide is used in coffee decaffeination and natural product extractions, as an alternative solvent in dry-cleaning and as a solvent in continuous flow hydrogenation reactions. An overview of these and some other industrial processes that use alternative solvents will be described in Chapter 10.

Unfortunately, as will become clear to readers, there is no universal green solvent and users must ascertain their best options based on prior chemistry, cost, environmental benefits and other factors. It is important to try and minimize the number of solvent changes in a chemical process and therefore, the importance of solvents in product purification, extraction and separation technologies has been highlighted.

There have been many in-depth books and reviews published in the area of green solvents. Hopefully, readers will find this book a readable introduction to the field. However, some cutting-edge results from the recent literature have been included in an attempt to give a clearer picture of where green solvents are today. For more comprehensive information on a particular solvent system, readers should look to the primary literature and the many excellent reviews of relevance to this field in journals such as Green Chemistry and Chemical Reviews.

Certain solvent media can be fascinating in their own right, not just as 'green' solvent alternatives! Therefore, we must not be blind to our overall goal in reducing the environmental burden of a particular process. Hopefully, readers of this book will be able to make up their own minds about the vast array of solvents available for a greener process, or even come up with a new addition for the green chemistry toolbox. Although many advances have been made during the past decade, the most exciting results are surely yet to come.

I would like to thank the editors of the RSC Green Chemistry Series, James Clark and George Kraus, for the opportunity to contribute a book to this important group of books. Also, I would like to acknowledge Merlin Fox (the commissioning editor) and the staff at RSC Publishing involved with this series, particularly, Annie Jacob, who has been advising and helping me all along the way. Finally, I would like to thank my husband, Chris Kozak, for his patience, support and motivational input during the writing of this book.

Francesca Kerton
St. John's, Newfoundland, Canada

Contents

RSC Green Chemistry Book Series
Alternative Solvents for Green Chemistry
By Francesca M. Kerton
© Francesca M. Kerton 2009
Published by the Royal Society of Chemistry, www.rsc.org

Chapter 3   Water

Chapter 4   Supercritical Fluids

Chapter 5   Renewable Solvents

Chapter 6   Room Temperature Ionic Liquids and Eutectic Mixtures

Chapter 7   Fluorous Solvents and Related Systems

Chapter 8   Liquid Polymers

CHAPTER 1

Introduction

1.1 The Need for Alternative Solvents

One of the 12 principles of green chemistry asks us to 'use safer solvents and auxiliaries'.[1-3] Solvent use also impacts some of the other principles and therefore, it is not surprising that over the last 10 years, chemistry research into the use of greener, alternative solvents has grown enormously.[4-8] If possible, the use of solvents should be avoided, or if they cannot be eliminated, we should try to use innocuous substances instead. In some cases, particularly in the manufacture of bulk chemicals, it is possible to use no added solvent—so-called 'solvent free' conditions. Yet in most cases, including specialty and pharmaceutical products, a solvent is required to assist in processing and transporting of materials. Alternative solvents suitable for green chemistry are those that have low toxicity, are easy to recycle, are inert and do not contaminate the product. There is no perfect green solvent that can apply to all situations and therefore decisions have to be made. The choices available to an environmentally concerned chemist are outlined in the following chapters. However, we must first consider the uses, hazards and properties of solvents in general.

Solvents are used in chemical processes to aid in mass and heat transfer, and to facilitate separations and purifications. They are also an important and often the primary component in cleaning agents, adhesives and coatings (paints, varnishes and stains). Solvents are often volatile organic compounds (VOCs) and are therefore a major environmental concern as they are able to form low-level ozone and smog through free radical air oxidation processes.[3] Also, they are often highly flammable and can cause a number of adverse health effects including eye irritation, headaches and allergic skin reactions to name just three. Additionally, some VOCs are also known or suspected carcinogens. For these and many other reasons, legislation and voluntary control measures have

RSC Green Chemistry Book Series
Alternative Solvents for Green Chemistry
By Francesca M. Kerton
© Francesca M. Kerton 2009
Published by the Royal Society of Chemistry, www.rsc.org

been introduced. For example, benzene is an excellent, unreactive solvent but it is genotoxic and a human carcinogen. In Europe, prior to 2000, gasoline (petrol) contained 5% benzene by volume but now the content is <1%. Dichloromethane or methylene chloride (CH_2Cl_2) is a suspected human carcinogen but is widely used in research laboratories for syntheses and extractions. It was previously used to extract caffeine from coffee, but now decaffeination is performed using supercritical carbon dioxide ($scCO_2$). Perchoroethylene (CCl_2CCl_2) is also a suspected human carcinogen and is the main solvent used in dry cleaning processes (85% of all solvents). It is also found in printing inks, white-out correction fluid and shoe polish. $ScCO_2$ and liquid carbon dioxide technologies have been developed for dry cleaning; however, such solvents could not be used in printing inks. Less toxic, renewable and biodegradable solvents such as ethyl lactate are therefore being considered by ink manufacturers.

Despite a stagnant period for the solvent industry during 1997–2002, world demand for solvents, including hydrocarbon and chlorinated types, is currently growing at approximately 2.3% per year and approaching 20 million tonnes annually. However, when the less environmentally friendly hydrocarbon and chlorinated types are excluded, market growth is around 4% per year. Therefore, it is clear that demand for hydrocarbon and chlorinated solvents is on a downward trend as a result of environmental regulations, with oxygenated and green solvents replacing them to a large extent.[9] It should be noted that these statistics exclude in-house recycled materials and these figures therefore just represent solvents new to the market; the real amount of solvent in use worldwide is far higher. It also means that annually a vast amount of solvent is released into the environment (atmosphere, water table or soil). Nevertheless the situation is moving in a positive direction, as in the USA and Western Europe environmental concerns have increased sales of water based paints and coatings to levels almost equal to the solvent based market. Therefore, it is clear that legislation and public interests are causing real changes in the world of solvents.

The introduction of legislation by the United States Food and Drug Administration (FDA) means that some solvents, *e.g.* benzene, are already banned in the pharmaceutical industry and others should only be used if unavoidable, *e.g.* toluene and hexane. FDA-preferred solvents include water, heptane, ethyl acetate, ethanol and *tert*-butyl methyl ether. Hexane, which is not preferred and is a hazardous air pollutant, is used in the extraction of a wide range of natural products and vegetable oils in the USA. According to the EPA Toxic Release Inventory, more than 20 million kg of hexane are released into the atmosphere each year through these processes. It may seem straightforward to substitute hexane by its higher homologue, heptane, when looking at physical and safety data for solvents (Table 1.1). However, heptane is more expensive and has a higher boiling point than hexane, so economically and in terms of energy consumption a switch is not that simple. Therefore, it is clear that much needs to be done to encourage the development and implementation of greener solvents.

Table 1.1 Properties of some volatile organic solvents, and some possible alternatives.

Solvent	Boiling point/°C	Flash point/°C	TLV–TWA^a/ppm	Hazards	Green?
Methanol	64	12	200	Toxic, flammable	Can be renewable
Ethanol	78	16	1000	Irritant, flammable	Can be renewable
Isopropanol	96	15	400	Irritant, flammable	
1-Butanol	117	12	100	Harmful, flammable	
Ethyl acetate	76	−2	400	Harmful, flammable	
Ethyl lactate	154	46	Not yet established	Irritant, flammable	Renewable
THF	65	−17	200	Irritant, flammable	
2-MeTHF	80	−11	Not yet established	Irritant, flammable	Renewable
2-Butanone	80	−3	200	Irritant, flammable	
Dichloromethane	40	none	100	Toxic, harmful, suspected carcinogen	
Chloroform	61	none	10	Possible carcinogen	
Toluene	110	4	50	Irritant, teratogen, flammable	
Hexane	68	−26	50	Irritant, reproductive hazard, flammable	
Heptane	98	−4	400	Irritant, flammable	
Water	100	none	Not applicable		Renewable, non-flammable, cheap
Carbon dioxide	Not applicable	none	5000	Compressed gas	Renewable, non-flammable, cheap
PEG-1000	Not applicable	none	Not applicable		Non-toxic, non-volatile
[Bmim] [PF$_6$]	Not applicable	none	Not yet established		Non-volatile

^{a}TLV–TWA: threshold limit value—time weighted average in vapour.

1.2 Safety Considerations, Life Cycle Assessment and Green Metrics

In recent years, efforts have been made to quantify or qualify the 'greenness' of a wide range of solvents; both green and common organic media were considered.[10,11] In deciding which solvent to use, a wide range of factors should be considered. Some are not directly related to a specific application, such as cost and safety, and these will generally rule out some options. For example, room temperature ionic liquids (RTILs) are much more expensive than water and they are therefore more likely to find applications in high value added areas such as pharmaceuticals or electronics than in the realm of bulk or commodity chemicals. However, a more detailed assessment of additional factors should be performed including a life cycle assessment, energy requirements and waste generation.

A computer-aided method of organic solvent selection for reactions has been developed.[12] In this collaborative study between chemical engineers and process chemists in the pharmaceutical industry, the solvents are selected using a rules based procedure where the estimated reaction–solvent properties and the solvent–environmental properties are used to guide the decision making process for organic reactions occurring in the liquid phase. These rules (Table 1.2) , whether computer-aided or not, could also be more widely used by all chemists in deciding whether to use a solvent and which solvents to try first.

The technique was used in four case studies; including the replacement of dichloromethane as a solvent in oxidation reactions of alcohols, which is an important area of green chemistry. 2-Pentanone, other ketones and some esters were suggested as suitable replacement solvents. At this point, the programme was not able to assess the effects of non-organic solvents because of a lack of available data. However, this approach holds promise for reactions where a VOC could be replaced with a far less hazardous, less toxic or bio-sourced option.

1.2.1 Environmental, Health and Safety (EHS) Properties

The EHS properties of a solvent include its ozone depletion potential, biodegradability, toxicity and flammability. Fischer and co-workers have developed

Table 1.2   Rules used in computer-aided solvent selection for organic reactions.

Establish need for solvents
Liquid phase reactions
The solvent must be liquid at room temperature
Need for solvent as carrier; if one or more reactants are solids
Need for solvents to remove reactants or products; if one or more products are solids
Need for phase split
Matching of solubility parameters of solute and solvent; within $\pm 5\%$ of the key reactant or product
Neutrality of solvents
Association/dissociation properties of solvents
EHS property constraints (based on up to 10 EHS parameters)

Table 1.3 Categories used in EHS assessment of solvents.

Release potential	Chronic toxicity
Fire/explosion	Persistency
Reaction/decomposition	Air hazard
Acute toxicity	Water hazard
Irritation	

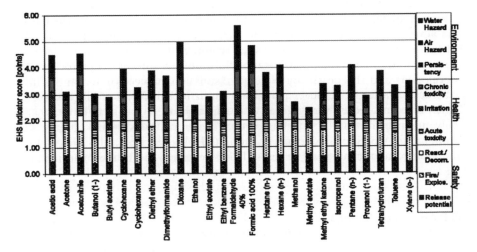

Figure 1.1 Results of an environmental, health and safety (EHS) assessment for 26 common solvents. [Reprinted with permission from *Green Chem.*, 2007, **9**, 927–934. Copyright 2007 The Royal Society of Chemistry.]

a chemical (and therefore, solvent) assessment method based on EHS criteria.[10] It is available at http://www.sust-chem.ethz.ch/tools/ehs/. They have demonstrated its use on 26 organic solvents in common use within the chemical industry. The substances were assessed based on their performance in nine categories (Table 1.3).

Using this EHS method, formaldehyde, dioxane, formic acid, acetonitrile and acetic acid have high (environmentally poor) scores (Figure 1.1). Formaldehyde has acute and chronic toxicity, dioxane is persistent and the acids are irritants. Methyl acetate, ethanol and methanol have low scores, indicating a lower hazard rating.

1.2.2 Life Cycle Assessment (LCA)

The function of life cycle assessment (LCA) is to evaluate environmental burdens of a product, process, or activity; quantify resource use and emissions; assess the environmental and human health impact; and evaluate and implement opportunities for improvements.[13] It is important to realize that while this book focuses on solvents, VOC 'free' paints and other 'green' consumer items may not be entirely green or entirely VOC free when the whole life cycle is considered. For example, a VOC may be used in the preparation of a pigment or another paint component, which is then incorporated into the final non-VOC (*e.g.* aqueous) formulation. The same can also be said for many synthetic procedures which are reported to be 'solvent free'. The reaction may be performed between neat reagents; however, a solvent is used in purifying, isolating and analysing the product. Chemists should be aware of this and avoid over-interpreting what authors are describing.

Fischer and co-workers undertook a LCA of the 26 organic solvents which they had already assessed in terms of EHS criteria (see above).[10] They used the Ecosolvent software tool (http://www.sust-chem.ethz.ch/tools/ecosolvent/), which on the basis of industrial data considers the 'birth' of the solvent (its petrochemical production) and its 'death' by either a distillation process or treatment in a hazardous waste incineration plant. For both types of end-of-life treatment, 'environmental credits' were granted where appropriate, *e.g.* solvent recovery and reuse upon distillation. The results of this assessment are shown in Figure 1.2. From an LCA perspective, tetrahydrofuran (THF), butyl acetate, cyclohexanone and 1-propanol are not good solvents. This is primarily due to the environmental

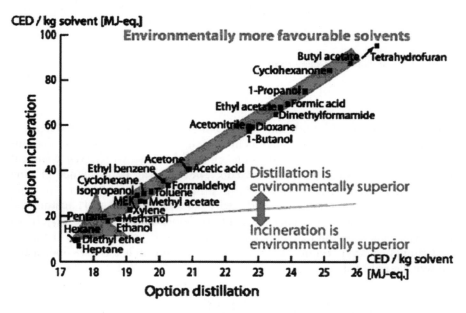

Figure 1.2   Life cycle assessment of the treatment options (incineration and distillation) for 26 common laboratory solvents. [Reprinted with permission from *Green Chem.*, 2007, **9**, 927–934. Copyright 2007 The Royal Society of Chemistry.]

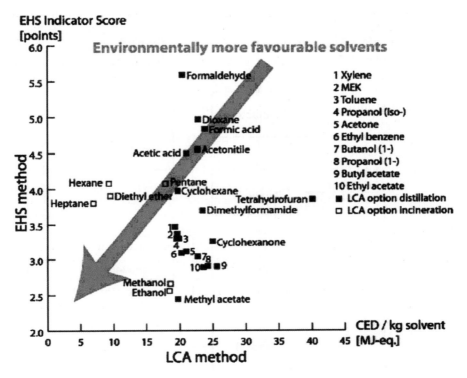

Figure 1.3 Combined EHS and LCA method for assessing 'greenness' of solvents. [Reprinted with permission from *Green Chem.*, 2007, **9**, 927–934. Copyright 2007 The Royal Society of Chemistry.]

impact of their petrochemical production and their LCA would therefore be better if they came from a different source. For example, 1-propanol may one day become available through selective dehydration and hydrogenation of glycerol (a renewable feedstock). At the other end of this scale, diethyl ether, hexane and heptane are considered favourable solvents. However, the reader should already be aware that diethyl ether is extremely hazardous in terms of flammability, low flash point and explosion risk through peroxide contamination. Therefore, the results from the EHS assessment and LCA were combined in an attempt to provide the whole picture (Figure 1.3).

It is evident from Figure 1.3 that formaldehyde, dioxane, organic acids, acetonitrile and THF are not desirable solvents. THF and formaldehyde are significant outliers on this last graph because of their particularly poor performance under one of the assessment methods. Methanol, ethanol and methyl acetate are preferred solvents based on their EHS assessment. Heptane, hexane and diethyl ether are preferred based on LCA. However, it must be noted that the LCA was performed based on petrochemical production of the solvents and if the first group of solvents was bio-sourced, perhaps these three solvents

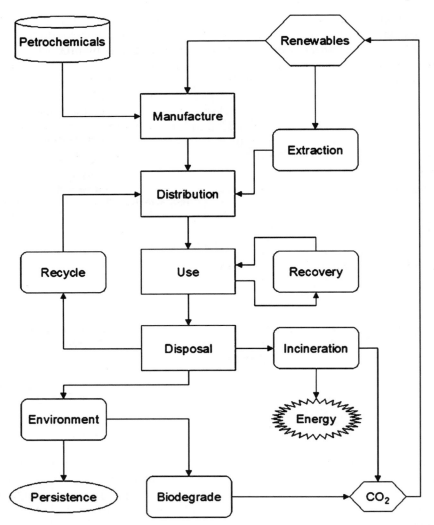

Figure 1.4   Life cycle flow chart for solvent usage. Primary life cycle stages are represented by rectangles. [Reprinted with permission from *Org. Proc. Dev.,* 2007, **11**, 149–155. Copyright 2007 American Chemical Society.]

would be the outright winners! Unfortunately, assessment tools used in this study could not be applied to many currently favoured alternative solvent technologies such as supercritical fluids and RTILs as there is a lack of available data at this time to quantify them fully.

However, a more qualitative LCA approach has been used by Clark and Tavener to assess the neoteric solvents described in this book (Figure 1.4).[11] The solvent must first be manufactured, usually from petroleum. This is relatively straightforward for simple and aromatic hydrocarbons that are obtained through cracking and distillation of crude oil. However, for other chemicals more complex synthetic routes are needed, *e.g.* to introduce heteroatoms such as halogens. Yet others, such as acetone, are produced as by-products in the manufacture of some chemicals. In terms of the alternative solvents described in this book, fluorous solvents and RTILs typically require multistage syntheses. Carbon dioxide and water do not need preparation but do need purification prior to use. Other renewable solvents, such as ethanol and esters, require separation or extraction and purification before use. A step often overlooked in LCA of chemicals is their distribution. Carbon dioxide and water are available globally and can therefore be sourced close to their point of use. Bioethanol would be a good solvent to use in Brazil but may not be readily available in other areas of the world. Therefore, the authors suggested a labelling system, similar to the 'food miles' being introduced at supermarkets, enabling chemists to find out where their compounds or solvents were manufactured.

The third primary stage in the life cycle of a solvent is its use. Solvents are used in many areas and not just as media for reactions (Table 1.4). The choice

Table 1.4 Some solvent applications.

Application	Description
Solvent extraction	In hydrometallurgy to recover metals from ores
	In nuclear fuel reprocessing
	In waste water treatment
	To recover natural products from plants or from fermentation liquors
	In organic synthesis and analytical chemistry
	As a degreaser and cleaning agent
Analytical chemistry and electrochemistry	Eluant in analytical and preparative chromatography, and in other separation techniques
	Dissolving the electrolyte to permit current to flow between the electrodes, without being oxidized or reduced itself
	As an oxidant or a reductant
Organic chemistry	As a reaction medium and diluent
	In separations and purification
	As a dehydrator (also in materials chemistry)
Polymer and materials chemistry	As a dispersant
	As a plasticizer
	As a blowing agent to create porosity
	As a binder to achieve cohesiveness in composite materials
	Production of powders, coatings, films, *etc.*
	As a developer in photoresist materials
Household and others	Fuels and lubricants
	Paints, varnishes, adhesives, dyes, *etc.*
	Antifreeze
	Cleaning fluids
	As a humectant (hydrating material) and in emulsions within cosmetics and pharmaceuticals

of the right solvent can have significant effects on energy consumption and the *E-factor* of a process. Solvent effects can lead to different reaction pathways for a number of reasons;[14] some of these effects will be briefly discussed later in this chapter. The E-factor is the mass ratio of waste to desired product.[15] If the wrong solvent is chosen, it can significantly affect the yield of a process (99% in the 'right' solvent compared to 30% in the 'wrong' one). For this reason, it is not surprising to find tables in journal articles showing the conversions or yields for a range of solvents. Clearly, in process development laboratories worldwide a significant amount of time and effort is spent optimizing the reaction conditions and the solvent choice to optimize this part of the LCA. Often the physical properties of the solvent play a significant role here; the boiling point and melting point, viscosity, volatility and density must all be considered alongside the safety issues such as flash point, reactivity and corrosiveness that were discussed earlier. At this stage in the process and the life cycle, biphasic systems and processes can be considered as these usually lead to reduced energy and increased efficiency.[11] Fluorous solvents can be advantageous for this reason. However, all alternative solvents have advantages and disadvantages. Unfortunately, in the chemical literature, most authors are biased and are trying to 'sell' their chosen reaction medium. For example, the pressures involved with supercritical fluids are a disadvantage, but the facile removal of the fluid at the end of a process is an advantage. Therefore, Clark and Tavener used a scoring system to grade the solvents (Table 1.5) in an attempt to qualify the general level of 'greenness' of a range of alternative solvents. It becomes apparent that all the solvents have some drawbacks and therefore solvent free approaches should attract greater attention. If a solvent is used, water should be considered first, and then carbon dioxide. They also suggest that it is unrealistic to think that all VOCs can be replaced in every application and therefore there is a growing role for VOCs derived from renewable resources in the alternative solvent field. In all areas, we need to balance the technical advantages of a particular solvent with any environmental, cost or other disadvantages.[3] For example, in the coatings industry, a reduction in the amount of VOC in paints may lead to a range of problems, including the stability of the formulation, longer drying times, a lower gloss and a less hard-wearing finish. However, aqueous emulsion paint has notable EHS advantages, including reduced VOC emissions, reduced user exposure and less hazardous waste production. Manufacturers and consumers need to decide if the advantages outweigh the disadvantages.

At the end of their life, solvents can often be reused or recycled by a range of recovery methods including distillation or biphasic separation. An environmental assessment of waste solvent distillation was recently reported and took into account a range of inputs and outputs including electricity consumption, cooling water, amount of recovered distillate and waste.[16] On average per kg of waste solvent processed, 0.71 kg of solvent is recovered, 1.4 kg steam, 0.03 kW h electricity, $1.5 \times 10^{-3} \, m^3$ nitrogen gas and $2.7 \times 10^{-2} \, m^3$ cooling water used. Steam is used for heating the waste solvent and nitrogen is used to avoid the formation of explosive vapour. Despite extensive recycling of solvents

Table 1.5 Advantages and disadvantages for alternative solvents, grades 1 (poor) and 5 (very good) for five different categories to give a maximum overall score of 25.[11]

Key solvent properties	Ease of separation and reuse	Health and safety	Cost of use	Cradle-to-grave environmental impact	Overall score /25
scCO₂					
Poor solvent for many compounds; may be improved with co-solvents or surfactants (1)	Excellent: facile, efficient, and selective (5)	Non-toxic; high-pressure reactors required (4)	Energy cost is high; special reactors; CO₂ is cheap and abundant (3)	Sustainable and globally available; no significant end-of-life concerns (5)	18
RTILs					
Designer/tailor-made properties; always polar (4)	Easy to remove volatile products; others may be difficult; reuse may depend on purity (2)	Limited data available; some are flammable and/or toxic (2)	Expensive; but low-cost versions may become available in time (2)	Mainly sourced from petroleum but some sustainable variants exist; synthesis may be wasteful and energy intensive; environmental fate not well understood (3)	13
Fluorous media					
Very non-polar solutes only; best used in biphasic systems (3)	Readily forms biphases; may be distilled and reused (4)	Bioaccumulative, greenhouse gases; perfluoropolyethers thought to be less problematic (2)	Very expensive (1)	Very resource demanding; may persist in environment	12
Water					
Possible to dissolve at least very small quantities of many compounds; generally poor for non-polar (3)	May be separated from most organics; purification may be energy demanding (3)	Non-toxic, non-flammable and safe to handle (5)	Very low cost; energy costs high (4)	Sustainable and safe to the environment; may need purification (4)	19
Bio-sourced solvents					
Wide range: ethers, esters, alcohols and acids are available (4)	May be distilled (4)	Generally low toxicity, can be flammable (4)	Mixed costs; will decrease with greater market volume and through biotech advances (4)	Sustainable resources, biodegradable, VOCs will cause problems (3)	19

within the chemical industry, ultimately the solvent will likely be incinerated at the end of its life (Figure 1.4). Incineration can generate valuable energy but the exhaust gases from the incineration plant also need treating.

Unfortunately, accidents happen; solvents can leak or spill and may not make it through to the normal end of their life cycle. Therefore, this possibility of release into the environment must also be taken into consideration when performing LCA. In these end-of-life scenarios, carbon dioxide has little environmental impact but other green solvents do. Water can become contaminated and must be treated prior to release. Fluorous solvents are difficult to incinerate and may form dangerous acidic by-products, and they are also persistent in the environment. However, perfluoroalkyl ether compounds, which have many similar properties to perfluorocarbons, are more short-lived in the environment and are therefore better solvents in terms of LCA for fluorous biphasic approaches. Unsurprisingly, new RTILs are being developed that take into account this part of a LCA and they are being designed with biodegradation in mind.[17,18]

1.2.3 Solvents in the Pharmaceutical Industry and Immediate Alternatives to Common Laboratory Solvents

The pharmaceutical industry is playing an active role in the development of green chemistry. In this industry solvents are a major concern and can have significant effects on the outcomes of complicated, multi-step synthetic procedures.[19–22] Although because of the smaller scale of pharmaceutical manufacturing—typically only 10–1000 tonnes of a product per year—the absolute amount of waste formed is quite low. Nonetheless, on a per kg basis the environmental burden of pharmaceutical synthesis is very high, as can be seen from the average E-factors (Table 1.6). Therefore, important advances have been made by this sector to employ the principles of green chemistry, and at the same time this reduces costs and increases profits for particular processes.

Solvent use accounts for 80–90% of mass utilization in a typical pharmaceutical batch chemical operation.[20] Additionally, solvents play a dominant role in the overall toxicity profile of most processes and are therefore the chemicals of greatest concern to many process development chemists. Table 1.7 lists the typical solvents used in a pharmaceutical process. It should also be noted that because of the multi-step procedures involved, an average of six different solvents are used in the manufacture of one bioactive compound. The avoidance of such multi-solvent approaches can have a significant impact on the amount of waste generated and overall productivity. For example, when the pharmaceutical company Pfizer redesigned their manufacturing process for the antidepressant drug sertraline, a three-step sequence was streamlined to a single step using ethanol as the only solvent.[23] This eliminated the need for dichloromethane, THF, toluene and hexane which had been used in the original process.

The table of solvents recently used by GlaxoSmithKline (GSK) shows a downward trend in the use of THF, toluene and dichloromethane. Additionally,

Table 1.6 E-factors in the chemical industry.[15]

Industry segment	Product tonnage	E-factor (kg waste : kg product)
Oil refining	10^6–10^8	<0.1
Bulk chemicals	10^4–10^8	<1–5
Fine chemicals	10^2–10^4	5–50
Pharmaceuticals	10–10^3	25–100

Table 1.7 Comparison of solvent use in GlaxoSmithKline Pharmaceuticals (GSK) over the last 15 years.[20]

	2005 rank[a]	1990–2000 rank
2-Propanol	1	5
Ethyl acetate	2	4
Methanol	3	6
Denatured ethanol	4	8
n-Heptane	5	12
THF	6	2
Toluene	7	1
Dichloromethane	8	3
Acetic acid	9	11
Acetonitrile	10	14

[a]Top 10 solvents used in GSK pilot plant processes during 2005.

it is reported that dichloromethane is the largest contributor to GSK materials of concern and there is an urgent need to develop alternatives for this solvent, or to develop different styles of reactor (not conventional batch reactors) which could reduce the amount of solvent required.

Pfizer have developed a solvent selection tool, which has been used to educate researchers about solvent replacement and has resulted in reduced amounts of chlorinated and ethereal solvents being used in their research labs.[19] A reduced availability of less desirable solvents also encouraged the uptake of alternatives. For example, hexane was replaced by heptane in stockrooms. The chart shown in Figure 1.5 could be applied to other industries and is easily used in academic research labs. It has been modified to take into account the findings of Fischer and co-workers, and as a result acetonitrile and THF have been transferred from usable to undesirable based on their performance in LCA.

The solvents in the 'black' category are there for a number of reasons: pentane and diethyl ether because of their low flash points; the chlorinated solvents, pyridine and benzene because they are carcinogens; and the polar aprotic solvents dimethylamine (DMA), *N,N*-dimethylformamide (DMF) and *N*-methyl pyrolidin-2-one (NMP) because they are toxic. Alternatives for many of these are readily available in most laboratories and some of them are listed in Table 1.8.

Unfortunately, no truly suitable alternatives to DMF, NMP and DMA are available at this time. Acetonitrile can be used in some cases but is not an ideal replacement.

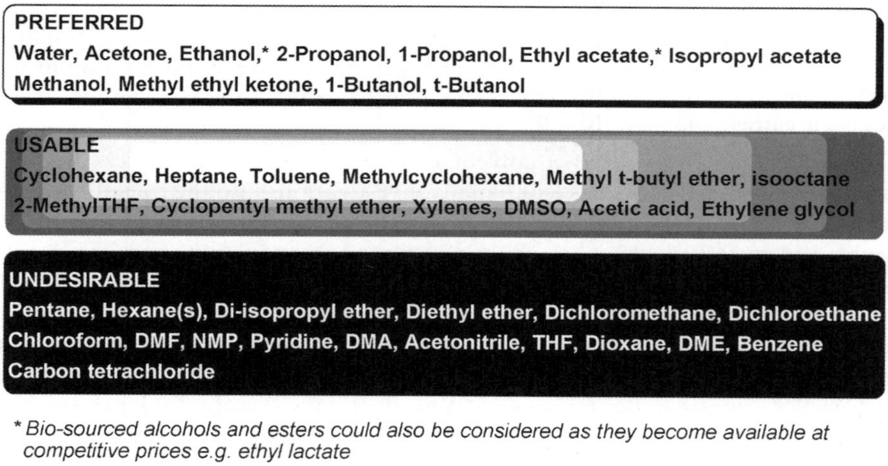

PREFERRED

Water, Acetone, Ethanol,* 2-Propanol, 1-Propanol, Ethyl acetate,* Isopropyl acetate
Methanol, Methyl ethyl ketone, 1-Butanol, t-Butanol

USABLE

Cyclohexane, Heptane, Toluene, Methylcyclohexane, Methyl t-butyl ether, isooctane
2-MethylTHF, Cyclopentyl methyl ether, Xylenes, DMSO, Acetic acid, Ethylene glycol

---

UNDESIRABLE

Pentane, Hexane(s), Di-isopropyl ether, Diethyl ether, Dichloromethane, Dichloroethane
Chloroform, DMF, NMP, Pyridine, DMA, Acetonitrile, THF, Dioxane, DME, Benzene
Carbon tetrachloride

---

* Bio-sourced alcohols and esters could also be considered as they become available at competitive prices e.g. ethyl lactate

Figure 1.5 Modified solvent selection guide.

Table 1.8 Possible alternatives for some 'blacklisted' solvents.[19]

Undesirable solvent	Alternative
Pentane or hexane(s)	Heptane
Ethers	2-MeTHF or methyl *t*-butyl ether (MTBE)
Dichloromethane (extractions)	Ethyl acetate, MTBE, toluene, 2-MeTHF
Dichloromethane (chromatography)	Ethyl acetate–heptane mixture

In 2005, the American Chemical Society Green Chemistry Institute (ACS GCI) Pharmaceutical Roundtable was established to encourage innovation and the uptake of green chemistry principles in this industry. It developed a list of key research areas, several of which were directly related to solvents.[21] These included solventless reactor cleaning, replacements for polar aprotic solvents (including NMP and DMF) and alternatives to chlorinated solvents for oxidations or epoxidations. The need to replace polar aprotic solvents is due to their designation as reproductive toxins and the resulting legislation that is coming into force.[20] Also, the mixed organic–aqueous waste that results from processes using these solvents is difficult to purify or incinerate.

1.3 Solvent Properties Including Polarity

Solvents can have a significant effect on the outcome of chemical reactions and physical chemical processes including extractions and crystallizations. Both the macroscopic (boiling point, density) and microscopic (dipole moment, hydrogen bonding ability) properties of the solvent affect its influence on such processes and the choice of solvent for a chemical system. For most paints and inks a volatile solvent is required, so a solvent is chosen with a relatively low boiling point and high vapour pressure. In reaction chemistry, the solvent plays many roles; it can act solely as the medium or it can participate in the reaction itself. For example, it can stabilize intermediates and increase rates of reaction; it can cause a shift in the equilibrium of a process; it can act as an acid or a base. There are many ways that a solvent can be involved in a process beyond solubilizing species. This has led to many investigations into the role of solvents in chemical reactions and further information can be found in textbooks published in that field.[14,24] A short review and introduction to solvents and solvent effects is also available.[25]

Although solvents are used as dispersing agents and in the formation of emulsions, they are generally used to dissolve materials. Whether this is to clean a surface or a reaction vessel, or to act as a heat transfer medium in a reaction, we need to consider the question—'Why do things dissolve?' Generally, the reasons are thermodynamic, in that if the dissolution process is energetically favourable it will occur. However, kinetics can also play a role and solutes that are poorly soluble at room temperature can be heated to increase solubility, a technique that is widely employed in recrystallizations.

Ionic compounds will dissolve in water if the Gibbs energy of solution (ΔG_s) is negative. As enthalpies of solution (ΔH_s) are usually negative and $\Delta G_s = \Delta H_s - T\Delta S_s$, most ionic species will dissolve. However, this does not mean that entropy of solution does not take a role. For non-ionic compounds to dissolve in a solvent, the Gibbs energy of mixing (ΔG_{mix}) must be negative (Figure 1.6). This can be encouraged by (1) the formation of strong intermolecular

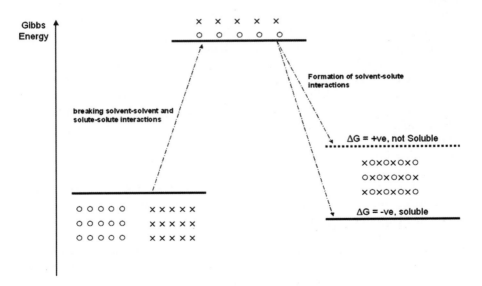

Figure 1.6 Simplified thermodynamic cycle for the dissolution of a compound ($\times$) in a solvent ($\bigcirc$).

interactions between the solute and the solvent, such as dipole–dipole interactions, or (2) the presence of very weak intermolecular forces in the solvent
and/or solute itself. Further explanation of this can be found in Chapter 4,
where the importance of such effects is crucial in utilizing the relatively poor
solvent power of supercritical carbon dioxide (scCO$_2$).

In general, one can use the rule 'like dissolves like' to qualitatively understand
solubility and solvent miscibility. Ionic compounds do not dissolve well in non-
polar solvents such as hydrocarbons, but generally will dissolve in many ionic
liquids. Ethanol, which is a polar molecule and capable of hydrogen bonding, is
miscible with water whereas hydrocarbons are immiscible. In fact, many of the
macroscopic physical properties of the solvent are dependent on the molecular
structure of the solvent molecules. For example, hydrogen bonding solvents
often have high heat capacities and this can be useful in the role of the solvent as
a heat transfer medium. Halogenated solvents typically have high densities and
this means they are useful in separations of aqueous–organic mixtures.
Throughout this book, selected physical parameters are given for some solvents.
However, further data can be found in the *CRC Handbook* and on materials
safety data sheets (MSDS) for most compounds.

There are many parameters that have been used to describe the attractive
forces (dispersive, dipolar and hydrogen bonding) present within a solvent or
liquid. However, *Hildebrand's solubility parameter* (δ) is probably the most
commonly used. In general, two liquids are miscible if the difference in δ is less
than 3.4 units. Also, if a solid (*e.g.* a polymer) has a δ similar to the solvent, it
will dissolve. However, there are exceptions to this rule especially with polar
solvents and solutes. Therefore, it is often worth testing solubility or solvent
miscibility on a small scale even if data are available.

'Polarity' is often used to predict the solubility of compounds, but unfortunately the concept is not straightforward. We all know that water and alcohols
are more polar solvents than hydrocarbons such as toluene and hexane.
However, we would not feel so confident describing the differences between
halogenated solvents and ethers in terms of polarity. This is because polarity is
actually described by several parameters and whether one solvent is more polar
than another often depends on which scale or parameter you are using. Solvent
polarity might best be defined as the *solvation power* of a solvent. It depends on
the interplay of electrostatic, inductive, dispersive, charge-transfer and hydrogen bonding forces.[4]

The terms polar, apolar and dipolar are often used to describe solvents and
other molecules, but there is a certain amount of confusion and inconsistency in
their application. *Dipolar* is used to describe molecules with a permanent dipole
moment, *e.g.* ethanol and chloroform. *Apolar* should be used rarely and only to
describe solvents with a spherical charge distribution such as supercritical
xenon. All other solvents should, strictly speaking, be considered *polar!*
Therefore, hexane is polar because it is not spherical and may be polarized in an
electric field. This polarizability is important when explaining the properties of
such solvents, which do not have a permanent dipole and give low values on
most polarity scales. Therefore, they are widely termed *non-polar* and, although

misleading, this term is useful in distinguishing solvents of low polarity from those with permanent dipoles.[4] Solvents that are able to donate an acidic hydrogen to form a hydrogen bond are termed *protic* (*e.g.* alcohols) and those that cannot are called *aprotic* (*e.g.* dimethyl sulfoxide).

Despite the problems of quantifying solvent polarity, numerous methods have been devised to assess polarity based on various physical and chemical properties. These include dielectric constant, electron pair acceptor and donor ability, and the ability to stabilize charge separation in an indicator dye. Many studies have been performed to assess the polarity of alternative solvents for green chemistry. The results are summarized in Figure 1.7.[11]

In addition to the terms defined in Table 1.9, empirical polarity scales have been developed based on *solvatochroism*. The most common solvatochromic dye used in these experiments is Reichardt's betaine dye (Figure 1.8). The UV-Vis spectrum of a solvatochromic dye changes in different solvents. In some cases, the dyes are modified to increase solubility for experiments in lower polarity media or alternative solvents.

The $\pi \rightarrow \pi^*$ transition for this dye varies between 810 nm (147 kJ mol^{-1}) and 453 nm (264 kJ mol^{-1}) on going from non-polar diphenyl ether to polar water. Polar solvents stabilize the zwitterionic ground state of the dye. This increases the energy difference between the π and π^* energy levels and leads to a higher energy (shorter wavelength) absorption. Values from these experiments are reported on a $E_T(30)$ scale, which reports the energy in kcal mol^{-1}, or on a E^N_T

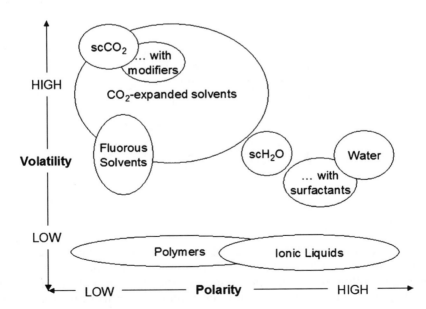

Figure 1.7 Typical polarity and volatility characteristics of alternative reaction media. [Reprinted with modifications and permission from *Org. Proc. Dev.*, 2007, **11**, 149–155. Copyright 2007 American Chemical Society.]

Table 1.9   Terms related to solvent polarity.

Term	Meaning/Definition
Dipole moment	Possessed by any compound with a non-symmetrical distribution of charge or electron density. Symmetrical molecules have no permanent dipole moment.
Dispersive forces	Weak intermolecular attractions as a result of instantaneous dipole–instantaneous dipole interactions.
Dielectric constant (ε_r)	Also known as relative permittivity, as it is measured relative to a vacuum. Measured by applying an electric field across the solvent (or vacuum) within a capacitor which induces a dipole in the solvent molecules and, therefore, takes into account polarizability. H_2O, 78.3; EtOH, 24.6; acetone, 20.6; toluene, 2.4; hexane, 1.9
Donor number (DN)	Measure of the Lewis basicity of a solvent. H_2O, 1.46; EtOH, 0.82; acetone, 0.44; toluene, <0.01; hexane, 0
Acceptor number (AN)	Measure of the Lewis acidity of a solvent. H_2O, 54.8; EtOH, 37.9; acetone, 12.5; hexane, 0

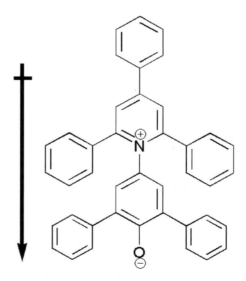

Figure 1.8   Negatively solvatochromic standard pyridinium-*N*-phenolate betaine dye, known as Reichardt's dye.

scale, which is normalized to reflect SI units (Table 1.10). On this scale, tetramethylsilane has a E^N_T value of 0.000 and water a value of 1.000. However, there are limitations to this procedure, as the dye molecule used is only reflecting limited types of molecular interaction based on its own structure. To overcome these limitations, the Kamlet–Taft parameters (α, β and π^*) were introduced and used for a series of seven different dyes. For each solvent, α is the hydrogen bond donor (HBD) ability, β is the hydrogen bond acceptor

Table 1.10 Polarity measurements for some solvents.

Solvent	$E_T(30)/kcal\,mol^{-1}$	$E^N{}_T$	α	β	π^*
Water	63.1	1.000 (defined)	1.17	0.47	1.09
Methanol	55.4	0.762	0.98	0.66	0.60
Ethanol	51.9	0.654	0.86	0.75	0.54
Acetone	42.2	0.355	0.08	0.43	0.71
Ethyl acetate	38.1	0.228	0.00	0.45	0.55
THF	37.4	0.207	0.00	0.55	0.58
Toluene	33.9	0.099	0.00	0.11	0.54
Hexane	31.0	0.009	0.00	0.00	−0.04

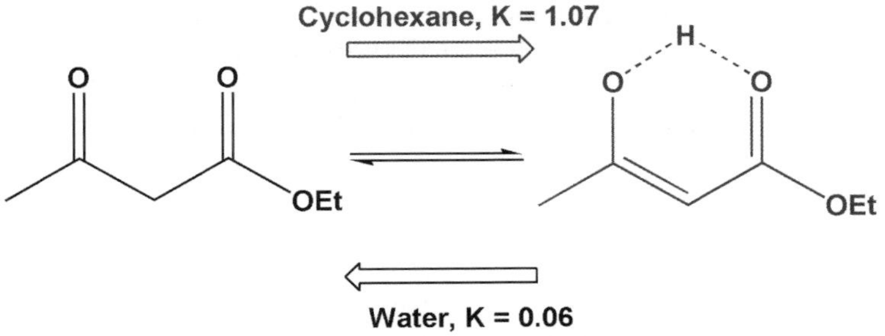

Figure 1.9 Solvent polarity effect on a keto–enol tautomerization.

(HBA) ability, and π^* is a measure of general polarity or polarizability (excluding hydrogen bonding effects). Obviously, to obtain this level of information involves a more complex procedure and Kamlet–Taft values are therefore often unavailable for alternative solvents.

The effect of solvent polarity on chemical systems including reaction rates and equilibria can be quite significant. In general, it is necessary to consider the relative polarities of the reactants and products. In equilibria, a polar solvent will favour the more polar species. A good example is the keto–enol tautomerization of ethyl acetoacetate shown in Figure 1.9. The keto tautomer is more polar than the enol tautomer and therefore the equilibrium lies to the left in polar media such as water Table 1.11.

A classic example of solvent polarity effects on reaction rates can be seen by comparing S_N1 and S_N2 nucleophilic substitution reactions. In an S_N1 reaction, an uncharged reactant (*e.g.* haloalkane) proceeds to a charged transition state and this will be stabilized by a more polar solvent, which will lead to a lowering of the activation energy and a faster reaction in a more polar solvent. However, in an S_N2 reaction the nucleophile is already charged and reacts with an uncharged substrate to give a transition state where the intermediate has a delocalized charge. This means that a polar solvent will stabilize the original, highly charged nucleophile and S_N2 reactions typically proceed more slowly in a more polar solvent than in a non-polar one. This is because of an increase in

Table 1.11 Some typical hydrogen bond donor (HBD) and hydrogen bond
acceptor (HBA) solvents.

HBD solvents	HBA solvents
Water	Acetonitrile
Acetic acid	THF
Methanol	Pyridine
Ethanol	Acetone
n-Propanol	(Water and alcohols)

the activation energy caused by stabilizing the reactant. In contrast, neutral reactants that pass through neutral intermediates on their way to neutral products generally show little change in reaction rate with changes in solvent polarity. However, as you will see later in this book, these are generalizations and sometimes enhanced reaction rates can occur unexpectedly when only neutral species are present throughout the whole reaction.

1.4 Summary

Green chemistry and the use of alternative solvents are intertwined. This is in part due to the hazards of many conventional solvents (*e.g.* toxicity, flammability) and the significant contribution that solvents make to the waste generated in any chemical process. Few solvents are inherently green, despite some misleading assertions in the literature. Although certain organic solvents are undesirable from both health and environmental points of view, most organic solvents can be handled safely in well designed plants with good recovery and recycle facilities. These plants should be able to adopt the new classes of bio-VOC solvents (Chapter 5) quite easily. However, increasingly, data and tools (from computer aids to simple tables or lists) are also available to ensure that if a VOC is chosen, it can be one with lower risks associated with it, *e.g.* heptane rather than hexane. Nevertheless, there are many alternatives to VOCs and although recent life cycle analyses suggest that some are greener than others, the choice of solvent really depends on the applications. New, often tailor-made media are being discovered on a regular basis which may be suitable, *e.g.* switchable solvents (Chapter 9). Alternative solvents have been developed and used for a wide range of properties. For example, in terms of volatility, we can choose from the most volatile supercritical carbon dioxide ($scCO_2$) to the least volatile polymeric and ionic liquid solvents. Volatility may be desirable in green chemistry in order to reduce the amount of residual solvent, or it may be undesirable with regard to atmospheric pollution. In terms of polarity, we can choose from polar aqueous phases to non-polar fluorous media. Accordingly, there should be a 'greener' solvent available for nearly every imaginable process and if there is not, it is just a matter of discovering it!

A wide range of reactions have been studied in many of the green alternative solvents that will be outlined in the following chapters. These can act as benchmarks for comparisons between solvent systems and include Diels–Alder (and retro-Diels–Alder) reactions, hydrogenations, hydroformylations, oxidations, carbon–carbon bond formations, polymerizations and metathesis reactions. However, surely the most exciting results are still to come in the manufacture of new 'benign-by-design' chemical products and materials which are yet to be imagined. Beyond reaction chemistry, there is even more to be discovered in the realm of alternative solvents. The application of green chemistry beyond the reaction (*e.g.* in analytical chemistry) is at a younger stage, so even more avenues are open for new, greener discoveries.

References

1. P. T. Anastas and J. C. Warner, *Green Chemistry: Theory and Practice*, Oxford University Press, New York, 1998.
2. J. H. Clark and D. J. Macquarrie, *Handbook of Green Chemistry and Technology*, Blackwell Science, London, 2002.
3. M. Lancaster, *Green Chemistry: An Introductory Text*, Royal Society of Chemistry, Cambridge, UK, 2002.
4. D. J. Adams, P. J. Dyson and S. J. Taverner, *Chemistry in Alternative Reaction Media*, John Wiley & Sons Ltd, Chichester, 2004.
5. K. Mikami, in *Green Reaction Media in Organic Synthesis*, Wiley-Blackwell, Oxford, 2005.
6. W. M. Nelson, *Green Solvents for Chemistry: Perspective and Practice*, Oxford University Press, Oxford, 2003.
7. D. Clarke, M. A. Ali, A. A. Clifford, A. Parratt, P. Rose, D. Schwinn, W. Bannwarth and C. M. Rayner, *Curr. Top. Med. Chem.*, 2004, **4**, 729.
8. R. A. Sheldon, *Green Chem.*, 2005, **7**, 267.
9. R. Hofer and J. Bigorra, *Green Chem.*, 2007, **9**, 203.
10. C. Capello, U. Fischer and K. Hungerbuhler, *Green Chem.*, 2007, **9**, 927.
11. J. H. Clark and S. J. Tavener, *Org. Process Res. Dev.*, 2007, **11**, 149.
12. R. Gani, C. Jimenez-Gonzalez and D. J. C. Constable, *Comput. Chem. Eng.*, 2005, **29**, 1661.
13. R. L. Lankey and P. T. Anastas, *Ind. Eng. Chem. Res.*, 2002, **41**, 4498.
14. C. Reichardt, *Solvents and Solvent Effects in Organic Chemistry*, Wiley-VCH, Weinheim, 2003.
15. R. A. Sheldon, *Green Chem.*, 2007, **9**, 1273.
16. C. Capello, S. Hellweg, B. Badertscher and K. Hungerbuhler, *Environ. Sci. Technol.*, 2005, **39**, 5885.
17. S. Bouquillon, T. Courant, D. Dean, N. Gathergood, S. Morrissey, B. Pegot, P. J. Scammells and R. D. Singer, *Aust. J. Chem.*, 2007, **60**, 843.
18. J. R. Harjani, R. D. Singer, M. T. Garcia and P. J. Scammells, *Green Chem.*, 2008, **10**, 436.
19. K. Alfonsi, J. Colberg, P. J. Dunn, T. Fevig, S. Jennings, T. A. Johnson, H. P. Kleine, C. Knight, M. A. Nagy, D. A. Perry and M. Stefaniak, *Green Chem.*, 2008, **10**, 31.

20. D. J. C. Constable, C. Jimenez-Gonzalez and R. K. Henderson, *Org. Process Res. Dev.*, 2007, **11**, 133.
21. D. J. C. Constable, P. J. Dunn, J. D. Hayler, G. R. Humphrey, J. L. Leazer, R. J. Linderman, K. Lorenz, J. Manley, B. A. Pearlman, A. Wells, A. Zaks and T. Y. Zhang, *Green Chem.*, 2007, **9**, 411.
22. A. M. Rouhi, *Chem. Eng. News*, 2002, **80**, 30.
23. G. P. Taber, D. M. Pfisterer and J. C. Colberg, *Org. Process Res. Dev.*, 2004, **8**, 385.
24. E. Buncel, R. Stairs and H. Wilson, *The Role of the Solvent in Chemical Reactions*, Oxford University Press, Oxford, 2003.
25. C. Reichardt, *Org. Process Res. Dev.*, 2007, **11**, 105.

CHAPTER 2
'Solvent free' Chemistry

2.1 Introduction

The greenest solvent, in terms of reducing waste, is no solvent.[1,2] Many industrial reactions are performed in the gas phase or without any solvent added. Similarly, many materials can be prepared without solvents via solid-state synthetic approaches. In addition to the term 'solvent free', the term 'solventless' is used in the literature to describe these reaction conditions. However, both these terms are somewhat misleading in many instances,[3] and the word 'neat' might be a better description to explain the highly concentrated nature of the reagents and lack of additional solvent. A solution is defined as a liquid mixture where the solute is uniformly distributed throughout the solvent. Therefore, whenever a solution is present, a solvent is also present. In many solvent free reactions (as this seems to be the most widely used term at this time) one of the reagents is a liquid and is sometimes present in excess. This liquid is often acting as the solvent and yielding a homogeneous reaction solution. In other solvent free reactions, there may be a liquid, *e.g.* water, formed during the course of the reaction and this liquid assists the reaction at the interface between the reagents and acts like a solvent. To add to the confusion, many reagents are commonly used in aqueous solution, *e.g.* 30% hydrogen peroxide. If no solvent is added to an oxidation reaction where the oxidant is aqueous hydrogen peroxide, is that an aqueous reaction or a solvent free reaction? If the reaction mixture is an emulsion or suspension of organic reagents in an aqueous phase, then the reaction is solvent free. However, if the reaction is occurring as a homogeneous solution, it is aqueous. Unfortunately, the situation is far more complex than this and is probably somewhere between the two extremes as many organic compounds are slightly miscible with water—even if only at a concentration of 0.1–1.0% by volume. Finally, in many solvent free approaches, VOCs are used to extract and purify the product, so although the reaction may be solvent free often the process as a whole does use solvents.

RSC Green Chemistry Book Series
Alternative Solvents for Green Chemistry
By Francesca M. Kerton
© Francesca M. Kerton 2009
Published by the Royal Society of Chemistry, www.rsc.org

Although a solvent is used in most of these 'solvent free' procedures, the amount of solvent required is dramatically reduced compared to conventional approaches and therefore these methods are generally very green. It should also be noted that some solvent free approaches lead to highly viscous solutions or indeed solid formation, in these cases, the technology may not be readily amenable for industrial development. However, recent results using ball mills for solvent free reactions are very promising. In this approach, a ball bearing is placed inside a vessel that is being shaken at high speeds. However, there are many variants of this approach, which are detailed in the introduction of a recent review.[4] The term *mechanochemistry* is sometimes used to describe this synthetic approach. It has already been successfully used for kilogram scale reactions, and details can be found in the recent review by Kaupp.[5] Industrially, similar results could possibly be achieved using cement mill technologies. It should be noted that the cement manufacturing is one of the largest scale chemical processes being conducted worldwide and that pilot-plant cement mills would normally be of a large enough scale and sufficiently well-engineered for many other chemical procedures. On a smaller scale in a research laboratory, solvent free procedures (unless otherwise indicated) use a mortar and pestle to grind the solid reagents together. Solvent free methods have also been very successfully utilized in combination with microwave heating.[6,7]

In some solvent free reactions, where the reaction involves a liquid, the process is similar to a conventional one except the conditions are highly concentrated. However, it should be noted that kinetic energy is supplied during the grinding of solid reagents and this can have several effects including heating and formation of surface defects.[8] Grinding also provides mass transfer and can prevent exothermic reactions forming hot spots, which would lead to decomposition. Mechanisms have been proposed, partially derived from atomic force microscopy data, regarding how solid–solid reactions of this kind proceed.[9] Initially, reagent molecules (A) migrate into cleavage planes or channels within the structure of the other reagent (B). The product (C) starts to form at the interface further distorting the crystalline structures, and a mixed A–B–C phase forms. Next, as the concentration of the product (C) increases, crystals of C begin to form within the A–B–C phase. In turn, the presence of growing amounts of C causes the mixed A–B–C phase to disintegrate and form new particles, which reveal fresh surfaces for further reaction.

Most of the recent literature in this field is concerned with synthetic organic reactions, supramolecular chemistry and crystal engineering. However, solvent free approaches can also be used in the extraction of natural products, although less information is available in the mainstream literature. Juice extractors can be used to afford aqueous solutions of biologically active compounds from undried plant material. An extract of *Capsicum annum* L. was recently prepared in this way, and then used in the green synthesis of silver nanoparticles.[10] The actual synthesis of the nanoparticles was conducted in the aqueous phase and therefore this work will not be discussed further here. However, this solvent free approach to extraction is probably worthy of greater representation in the green chemistry literature.

In the rest of this chapter, selected examples will be described and discussed; however, solvent free organic synthesis is a rapidly growing field and more examples can be found in recent reviews and a book on the subject.[1–2,4,6–8] In reaction schemes within this chapter, room temperature means the temperature at which the solventless procedure was set up. During the course of solvent free reactions (particularly in ball mills) heat is generated and therefore the reaction might proceed at a higher temperature even though no external heat is applied.

2.2 Chemical Examples

2.2.1 Inorganic and Materials Synthesis

In the area of renewable materials, bulk oxypropylation of chitin and chitosan has been performed.[11] Chitin and chitosan are abundant natural polymers obtained from shellfish, such as crab shell or shrimp shell. This solvent free reaction yields viscous polyols. Unfortunately, propylene oxide homopolymer is formed as a by-product but is easily separated. It should be noted that care was taken to minimize the risk involved in the use of toxic, flammable propylene oxide (the reagent in this process).

In the field of nanotechnology, limited solvents were recently employed in the microwave assisted synthesis of nickel–graphitic shell nanocrystals (Figure 2.1).[12] Nickel nanoparticles were blended with poly(styrene) (PS) using ethyl acetate and a sonicator. The solvent was removed under vacuum to yield a nickel-containing PS phase. When this was heated in a microwave under solvent free conditions, nickel–graphitic shell nanocrystals were formed. The nickel cores could be dissolved using 1 M hydrochloric acid and ultrasound to give hollow carbon nanospheres. Given this initial study, there is probably a lot of potential for solvent free microwave synthesis in the preparation of new nanomaterials.

Solvent free methods have been used extensively in supramolecular chemistry, coordination chemistry and the formation of transition metal clusters and polymers.[8] Reactions range from very simple ligand substitution reactions for salts of labile metal ions[13] to more complex procedures, some of which are outlined below.

Calix[4]resorcinarenes, which can be used as supramolecular building blocks, have been prepared in high yields and purity using a solvent free approach.[14] Equimolar quantities of the benzaldehyde and resorcinol, in the presence of *p*-toluenesulfonic acid, were ground together using a mortar and pestle

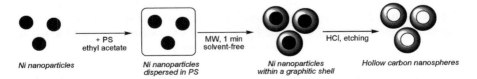

Figure 2.1 Preparation of graphitic carbon nanostructures using a microwave induced solid state process.

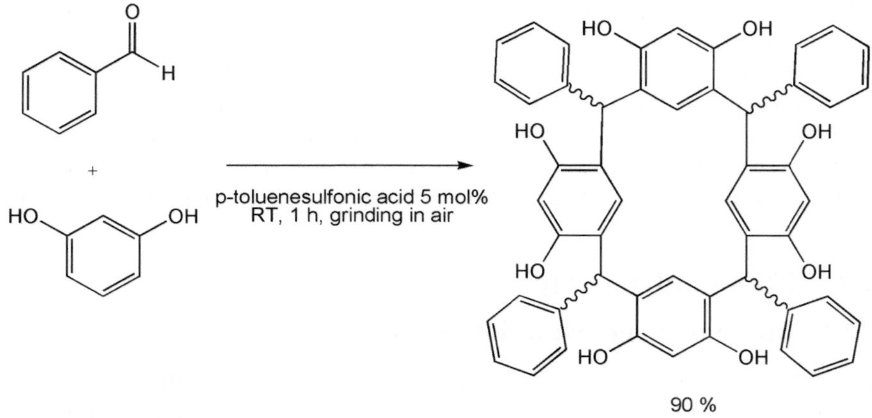

Figure 2.2 Solvent free synthesis of calix[4]resorcinarene.

(Figure 2.2). The reaction occurs between the two mutually dissolved reagents to afford a viscous paste that becomes red in colour upon standing for 1 h. The products could be easily purified using water to remove the acid and hot methanol for recrystallization.

Supramolecular self-assembly has been found to be dramatically accelerated in some cases, and this is perhaps the reason why solventless reactions have been rapidly adopted by coordination chemists. For example, the formation of a platinum-containing square is dramatically accelerated compared with the same reaction in water (Figure 2.3).[15] In water, the reaction needed to be heated at 100 °C for 4 weeks. In a solvent free approach, approximately the same yield was achieved in 10 min at room temperature. Water and ethanol were the only solvents used in the work up of the reaction. This approach was then extended to bowl-shaped and helical supramolecular structures.

One-dimensional coordination polymers of copper, zinc and silver have also been prepared using solvent free techniques (Figure 2.4).[16–18] This was achieved by grinding the ligand (DABCO or *trans*-1,4-diaminocyclohexane) with the metal precursor for 5 min, followed by recrystallization in water–methanol. No yields are reported for many of these reactions because of the small scale on which they were conducted. However, in some cases, different structures are reported when the reaction is conducted in solution compared with the solid state.

A reversible solid-state HCl elimination reaction from a Cu(II) pyridinium coordination complex has been reported.[19] The reaction proceeds with a colour change from yellow (pyridinium complex) to blue (pyridine complex). This reaction suggests that other protic ligands may successfully be coordinated to metals using a solvent free approach in the future.

In an interesting twist on the solvent free reaction, Petrukhina and co-workers co-sublimed a volatile metal complex, [Rh$_2$(CF$_3$CO$_2$)$_4$], and arene molecules including paracyclophanes to yield organometallic coordination polymers.[20,21] Moderate yields of crystalline products were obtained (35–70%),

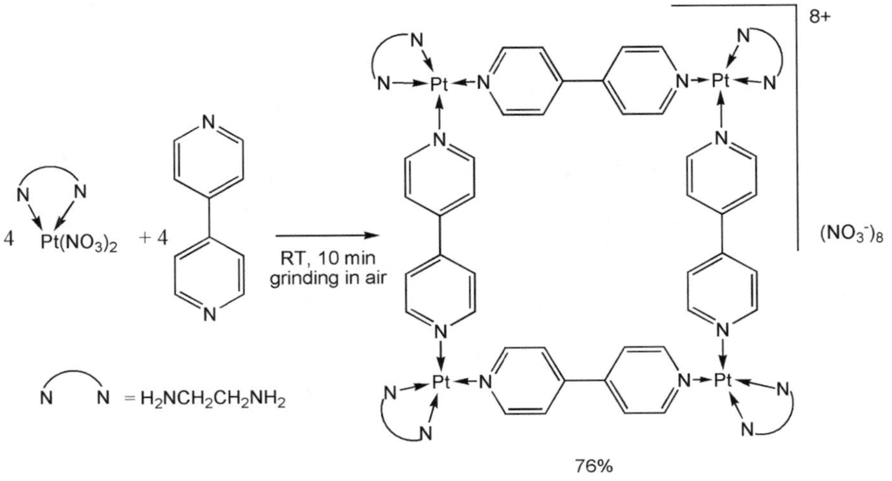

Figure 2.3 Solvent free supramolecular self assembly of a metallo-square.

and because of the highly porous structure of the material these may find applications in gas recognition and sorption. Therefore, the formation of interesting transition metal complexes is not restricted to solid–solid grinding procedures.

The use of solvent free conditions is also not limited to the preparation of complexes and materials containing transition metals. Molecular inorganic molecules containing main-group elements have also been prepared in this way, *e.g.* iodination of *ortho*-carboranes (Figure 2.5).[22] Yields are significantly higher than in procedures performed using chlorinated solvents. The gaseous hydrogen iodide by-product is removed by evaporation and the excess iodine (required to get near quantitative yields of the desired product) can be removed by sublimation. Pure tetraiodinated carborane, with potential uses as an radiographic contrast agent, can be obtained by recrystallization from ethanol–water. The use of solvent free procedures is becoming commonplace for many small-molecule main-group reagents that are prepared and used industrially, such as chlorosilanes, but there is still probably extensive scope for expanding the use of solvent free approaches in this field.

2.2.2 Organic Synthesis

Despite recent interest in solvent free procedures within inorganic and materials chemistry, most of the reactions studied in this way have been organic. Reactions that have been studied fall into two main classes: thermal and photochemical.[2] Thermal solvent free reactions to date include oxidations, reductions, isomerizations, additions, eliminations, substitutions, carbon–carbon couplings (including cycloadditions, condensations, Reformatsky, Wittig), pinacol couplings, phenol couplings, oxidative couplings (Glaser) and

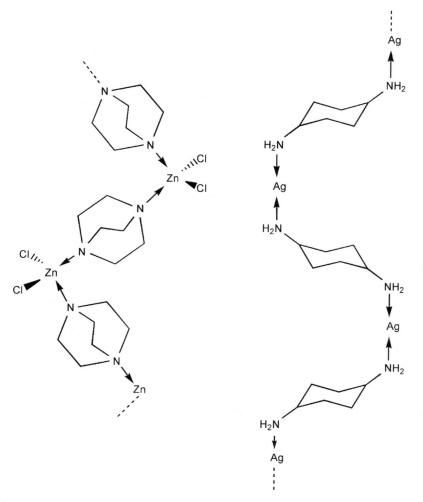

Figure 2.4 Some 1D-coordination polymers formed using solvent free methods: Zn(II)-DABCO (left) and Ag(I)-*trans*-1,4-diaminocyclohexane (right).

polymerizations. Photoreactions include dimerization, polymerization, cycliza-tion, isomerization, decarbonylation and addition. Therefore, many solvent free reactions have been studied (Figure 2.6) and reactions are perhaps more amenable for study in this way than many of us would at first realize. For instance, reactions are definitely more amenable to study under solvent free conditions than in supercritical carbon dioxide. However, some organic reac-tions proceed explosively in the solid state or under neat liquid conditions. In those cases, a solvent is required to mediate the reaction. However, as the long list of reactions that have been studied demonstrates, many reactions proceed moderately in the absence of solvent or in a water suspension and

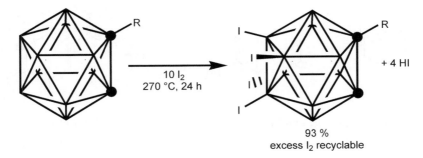

Figure 2.5 Solvent free iodination of *ortho*-carboranes.

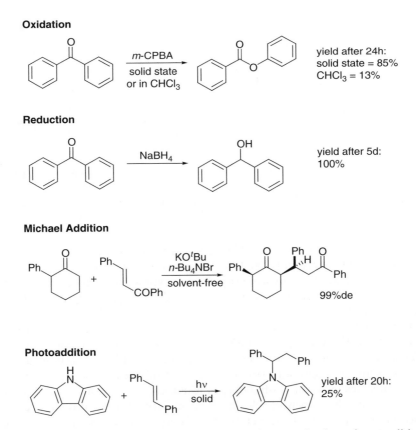

Figure 2.6 Some synthetic organic reactions that have used solvent free conditions.

therefore solvent free conditions should be employed wherever possible in an aim to reduce solvent usage and waste.

Dihydropyrimidinones, which have promising biological activities as anti-hypertensive and anticancer drugs, can be prepared through a solvent free

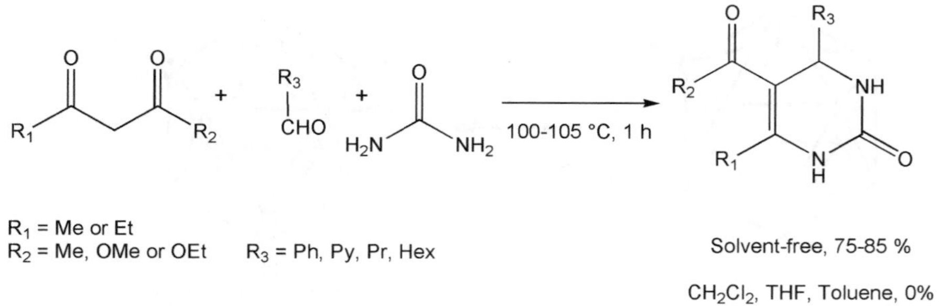
R1
R2
+
R3
CHO
+
H2N
NH2
100-105 °C, 1 h
R2
R3
NH
R1
N
H
O
R1 = Me or Et
R2 = Me, OMe or OEt
R3 = Ph, Py, Pr, Hex
Solvent-free, 75-85 %
CH2Cl2, THF, Toluene, 0%

Figure 2.7 Solvent and catalyst free Biginelli condensation of a 1,3-dicarbonyl compound, an aldehyde and urea.

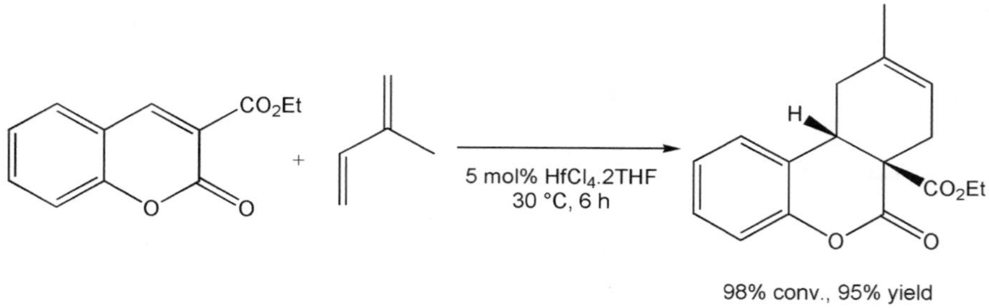
CO2Et
+
5 mol% HfCl4.2THF
30 °C, 6 h
H
CO2Et
98% conv., 95% yield

Figure 2.8 Solvent free hafnium chloride catalysed Diels–Alder cycloaddition of an ethoxycarbonylcoumarin with 1,3-diene.

Biginelli reaction (Figure 2.7).[23] The 1,3-dicarbonyl compound, aldehyde and urea are heated to around 100 °C for 1 h without any need for solvent or catalyst and yield the dihydropyrimidone in good to excellent yields. Performing the same reaction in refluxing VOCs did not yield any product, indicating a special advantage for a solventless procedure in this case. The reaction has been scaled up successfully to the 1 kg level and the only additional solvents required to work up the product effectively are water and ethanol.

Diels–Alder cycloadditions can be catalysed by $HfCl_4 \cdot 2THF$ in air and under solvent free conditions (Figure 2.8).[24] 3-Ethoxycarbonylcoumarins were successfully reacted with 1,3-butadienes to afford the cycloadduct in excellent yield. The tetrahydrobenzo[*c*]chromenone moiety of the product occurs in many natural products and therefore this reaction is of interest for the synthesis of biologically active compounds. Significantly, the yields are much higher under solvent free conditions than in conventional reaction media (Table 2.1). Reactions performed in water or in a typical ionic liquid were also less effective. It should also be noted that one of the reagents, isoprene (2-methyl-1,3-butadiene), is a liquid at room temperature and used in excess, so it may be playing a solvation role in this reaction. Additionally, ethyl acetate and petroleum ether

Table 2.1 Conversions for hafnium chloride catalysed cycloaddition reactions in different reaction media.[24]

Reaction medium	Conversion %
Solvent-free	98
Nitromethane	50
Dichloromethane	40
Acetonitrile	10
THF	2
H_2O (pH 1.6)	<1
[Bmim][PF_6]	35

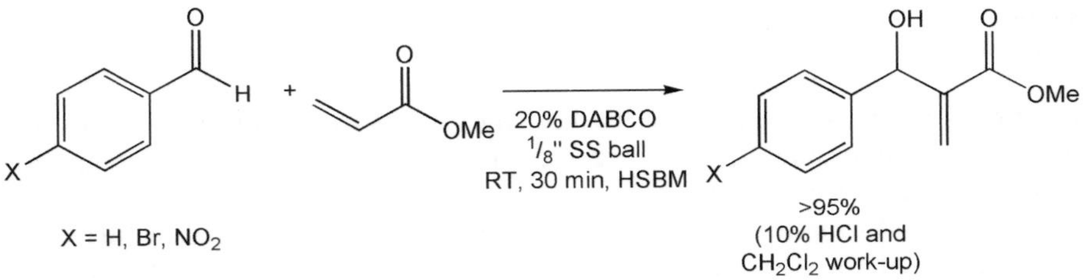

X = H, Br, NO_2

20% DABCO
$^1/_8$" SS ball
RT, 30 min, HSBM

>95%
(10% HCl and
CH_2Cl_2 work-up)

Figure 2.9 Base catalysed Baylis–Hillman reaction under solvent free conditions and using an HSBM.

were used to purify the product by column chromatography. Therefore, although reduced in solvent demand this reaction is not entirely solvent free.

Another atom-efficient process that has been studied solvent free is the Baylis–Hillman reaction.[25] This reaction affords useful multifunctional products from an addition reaction between an electrophile (often an aldehyde) and an electron-deficient olefin. Unfortunately, under most conditions it has the significant drawback of a slow rate of reaction. However, this has been overcome through a solvent free approach that uses a high-speed ball mill (HSBM) (Figure 2.9). Previous solvent free studies of this reaction took 3–4 days to achieve completion. In contrast, using an HSBM, the reaction is complete in 30 min. Unfortunately, a chlorinated solvent was chosen for reaction work up; clearly, it would be desirable to use a less hazardous VOC here.

Ball mills have been used for many carbon–carbon bond-forming reactions and this field was recently reviewed.[4] Reactions studied include catalyst-free processes such as Knoevenagel condensations, organic- and base catalysed reactions including the Baylis–Hillman reaction discussed above and asymmetric aldol reactions discussed below, and metal-mediated or catalysed reactions including the Suzuki and Heck reactions. In a ball mill, the reaction mixtures are known to heat up and generate a considerable pressure; therefore for the Heck reaction an attempt was made to ascertain the influence of both these parameters. Although conversions were obtained by compressing the

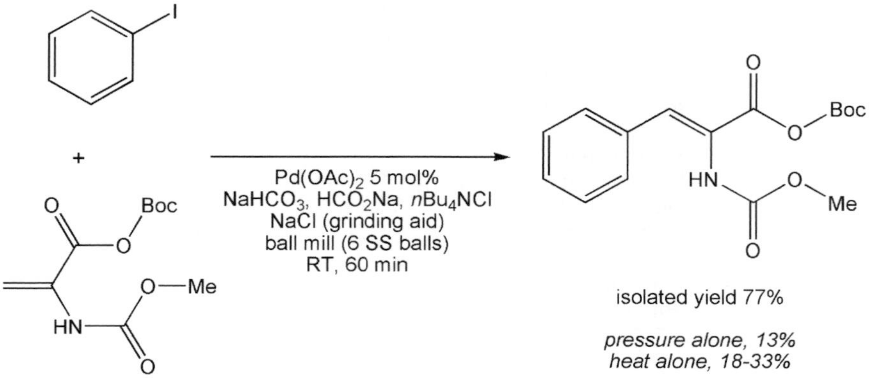

Figure 2.10   Solvent free Heck reaction.

reaction mixture in an anvil and by heating solvent free reaction mixtures in a test-tube, the combination of effects present in a ball mill was found to give far superior results (Figure 2.10).[26,27]

Many transformations of fullerenes have also been performed in this way.[4] The earliest of these was a [2 + 2] cycloaddition (Figure 2.11) that led to the formation of a dumb-bell shaped C_{120} molecule, probably through a fullerene radical anion intermediate.[28] If the same reaction was performed in the liquid phase, a cyanated product ($C_{60}HCN$) was formed. More recently it has been shown that the choice of base in a solvent free Bingel reaction is essential.[29] Under conventional conditions 1,8-diazabicyclo[5.4.0]undec-7-ene (DBU) is most commonly used as the base for this reaction; however, no cyclopropanated product was formed under solvent free conditions. Other organic bases, such as piperidine and triethylamine increased the yield somewhat, but the inorganic base sodium carbonate under optimized conditions gave the superior yield of 78% (Figure 2.11). This study highlights the importance of optimizing a solvent free process and that conditions which typically work in solvent do not necessarily directly convert to good solvent free procedures.

Gold- and platinum catalysed polycyclizations have also been performed under solvent free conditions (Figure 2.12).[30] These isomerization reactions, atom-efficient by their very nature, offer an expedient and stereoselective route to various polycyclic products. Although studies are ongoing, the catalyst loading can be reduced to 0.5% and in some cases the product can be isolated by distillation. In other cases, petroleum ether was used to purify the crude reaction mixture by column chromatography.

Gold has also been used to catalyse multi-component addition/condensation reactions in water and under solvent free conditions (Figure 2.13).[31] This atom-economic method rapidly gave a range of aminoindolizines in high yields with low catalyst loadings. The reaction was performed in water but yields were generally greater when the reaction was performed under solvent free conditions. In many cases, all three reagents are liquids and the product is purified

Fullerene [2+2] cycloaddition

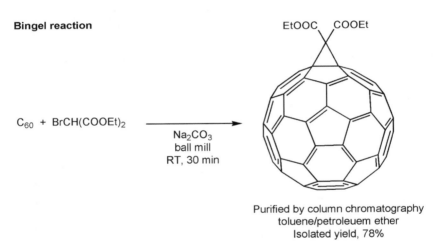

Figure 2.11   Some reactions of fullerenes performed under solvent free conditions in a ball mill.

using chromatography with the primary purification solvent being ethyl acetate, so solvent is present in the whole chemical process. The approach was extended to couplings using amino acid derivatives in place of the amine. In these cases, no loss of enantiomeric purity was lost. In summary, this route provides an excellent, low E-factor route to compounds with potential biological activity and pharmaceutical uses.

A range of organic transformations promoted by lithium bromide and triethylamine under neat reaction conditions have been reported.[32] As the reagents benzaldehyde and triethylamine are liquids, these reactions may not be entirely solvent free, just without an added solvent. The product distribution (or class of reaction) was affected by the solvent used in the reaction work up (Figure 2.14), and therefore a wide range of products can be obtained using a very simple approach.

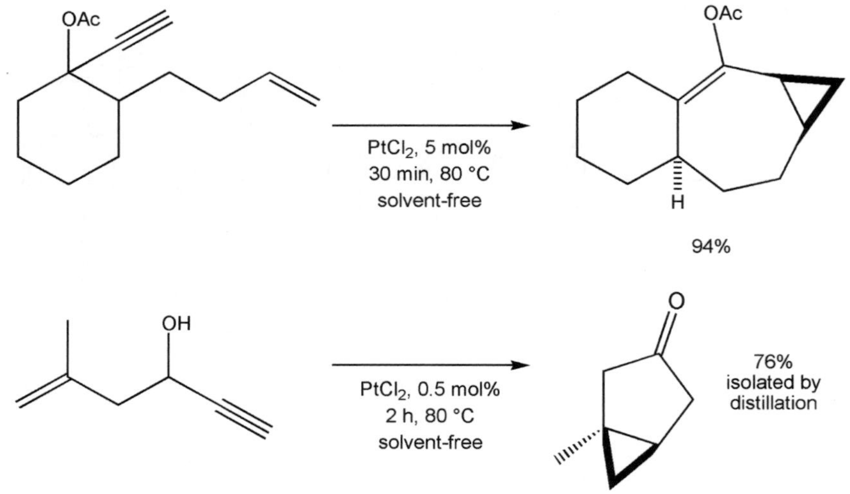

Figure 2.12 PtCl$_2$ catalysed cycloisomerization under neat reaction conditions.

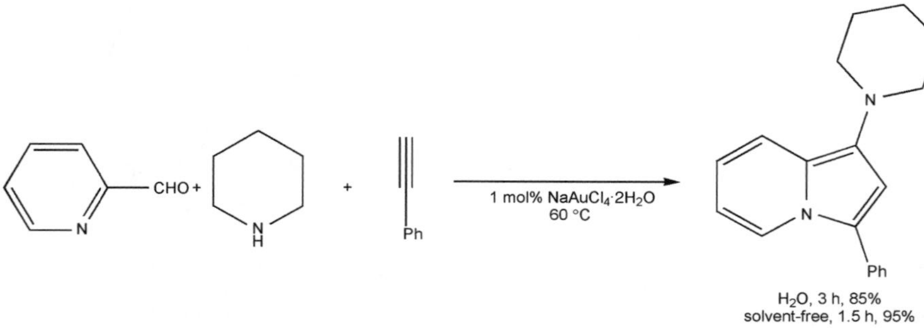

H$_2$O, 3 h, 85%
solvent-free, 1.5 h, 95%

Figure 2.13 Gold catalysed three-component coupling reaction in water and under solvent ree conditions.

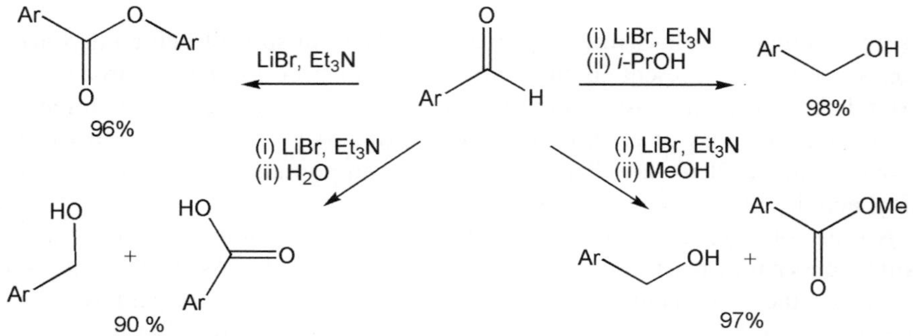

Figure 2.14 Solvent free Cannizzaro, Tischenko and Meerwein–Ponndorf–Verley reactions.

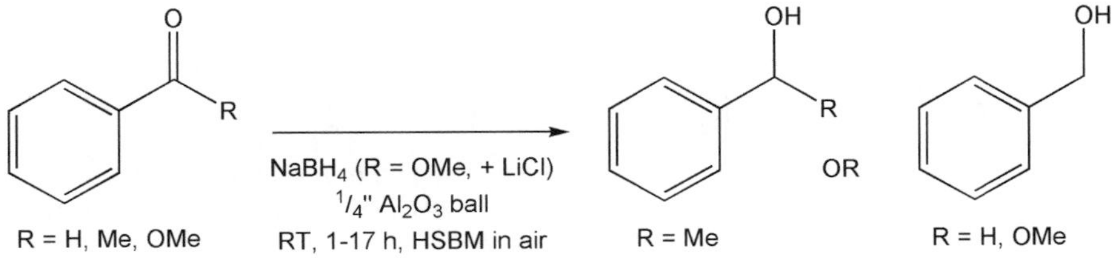

Figure 2.15   Solvent free reductions with sodium borohydride using an HSBM.

Reduction of substituted benzaldehydes, acetophenones and methyl benzoates has been performed under solvent free conditions (Figure 2.15).[33] Similar solvent free reductions had previously been reported, but these required grinding in a mortar and pestle for 5 days under an inert atmosphere. By performing the reaction in an HSBM, Mack and co-workers were able to perform reactions on an open bench in air and reaction times were reduced to between 1 h and 17 h. It should be noted that in one case, the reduction of *p*-nitrobenzaldehyde, the reaction was highly exothermic and yields and conversions could not be determined. Therefore, such methods should be used with some caution, as when no solvent is present reactions can suffer from the lack of a heat transfer medium. Importantly, in working up the reactions, only 10% aqueous hydrochloric acid and water were used to quench the reaction and purify the product. If solvent was required to aid purification, the relatively benign VOC methanol was used.

Thirty-nine different dithiocarbamates have been efficiently prepared through a one-pot reaction of an aliphatic primary or secondary amine, carbon disulfide and an alkyl halide (Figure 2.16).[34] Typically, all reagents were liquids and the mixture slowly solidified upon reaction. Therefore, although the reaction does not use solvents, the reagents are probably acting as the solvent in this procedure. Additionally, the reactions were quenched with water and extracted with ethyl acetate, so solvent was used. Nevertheless, this is an excellent synthetic method for the preparation of *S*-alkyl dithiocarbamates, which are useful compounds for the pharmaceutical and agrochemical industries.

Solvent free methods have also impacted on the preparation of other alternative reaction media. Namely, a range of ionic liquids (ILs) was prepared (including imidazolium, pyridinium and phosphonium salts) through halide-trapping anion metathesis reactions (Figure 2.17). The alkyl halide by-product was easily removed by vacuum or distillation and the products were obtained quantitatively in high purity. In addition to being solvent free, this route is more atom economic than the usual route to room temperature ionic liquids (RTILs) as it does not use silver(I), alkali metal or ammonium salts which are normally used in an anion metathesis reaction.

Related heterocyclic salts, imidazolinium chlorides, have recently been prepared through a solvent free reaction of a formamidine with dichloroethane and a base.[35] Solvents are used in this reaction as excess dichloroethane is used and the residue is triturated and washed using either acetone or toluene. However, it is a superior route to these valuable *N*-heterocyclic carbene

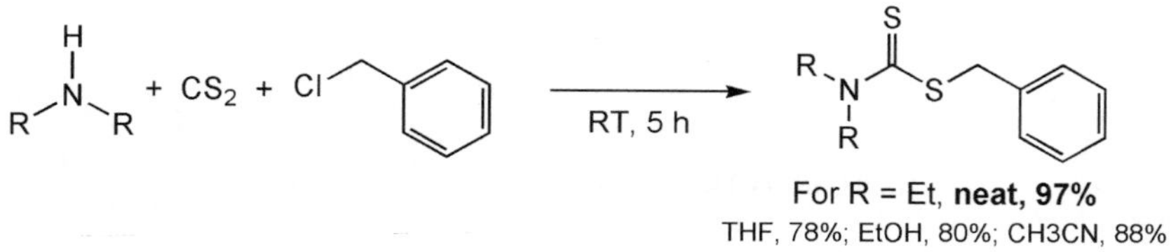

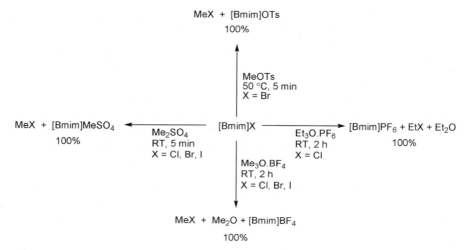

Figure 2.16 Catalyst free one-pot synthesis of dithiocarbamates.

Figure 2.17 Quantitative synthesis of 1-butyl-3-methylimidazolium [Bmim] ionic liquids via solvent free anion metathesis.

precursors, as previous methods often required careful chromatography of an unstable diamine intermediate.

Solvent free biocatalytic peptide syntheses have been successfully performed, as have esterification reactions. 'Solvent free' approaches in biocatalysis typically involve liquid substrates and either free or immobilized enzymes; however, solid–solid reactions have also been studied.[36] Because of the highly concentrated nature of the reagents, the initial reaction rates are generally high. Also, in solvent free polymerizations, rates can be significantly affected by chain entanglement and lack of access to the active site. In some examples, small amounts of co-solvents are used to decrease the melting point of solid reagents, so that temperatures are not unreasonable for thermally sensitive enzymes.

2.2.2.1 Enantioselective Catalysis

Many catalysts exhibit a decrease in enantioselectivity in the absence of solvent, but there are some examples where stereoselectivity actually increases.[37] Solvent free asymmetric catalysis was recently reviewed, and reactions studied to

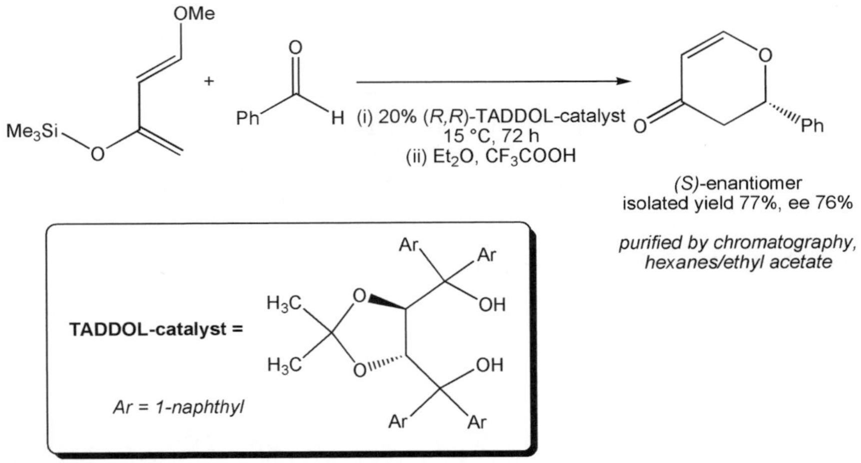

Figure 2.18 Catalytic asymmetric hetero-Diels–Alder reaction between Danishefsky's diene and benzaldehyde under solvent free conditions.

date include epoxide-opening reactions, hetero-Diels–Alder reactions, ring-closing metathesis reactions, Michael additions, hydrogenations, hydroformylations and conjugate additions.[37]

Solvent free conditions were used in a hetero-Diels–Alder reaction catalysed by a TADDOL ($\alpha,\alpha,\alpha',\alpha'$-tetraaryl-1,3-dioxolan- 4,5-dimethanol) derivative (Figure 2.18).[38] This class of catalyst was found to act through an intermolecular hydrogen bonding mechanism and it is therefore not surprising that enantioselectivities are enhanced under solvent free conditions or in the presence of a hydrogen bonding solvent such as ethanol. In this example, the reaction mixture was extracted using diethyl ether, trifluoroacetic acid was used to quench the reaction and the product was purified using column chromatography. Therefore, although the reaction is solvent free, there is still considerable room to improve the overall E-factor for the process.

Chiral diphosphine diamine complexes of ruthenium have been found to effectively catalyse the hydrogenation of ketones and imines in the presence of an alkoxide base under 3 atm hydrogen.[39] The reaction was performed solvent free (neat reagents) when the imine or ketone were liquids. However, benzene was used for solid reagents. In general, conversions were excellent (100% over a 4–12 h reaction period) and enantioselectivities were good to excellent. Carbonyl groups could be selectively reduced in the presence of C=C bonds. Importantly, very high substrate:catalyst ratios could be achieved using these neat reaction conditions, typically between 3000:1 and 5000:1. This means that a smaller amount of the expensive catalyst can be used and it also could lead to less metal contamination in the product.

Solvent free sulfide oxidation (Figure 2.19) has been performed using a chiral aluminium(salalen) complex.[40] Enantioselectivity and yields were found to be

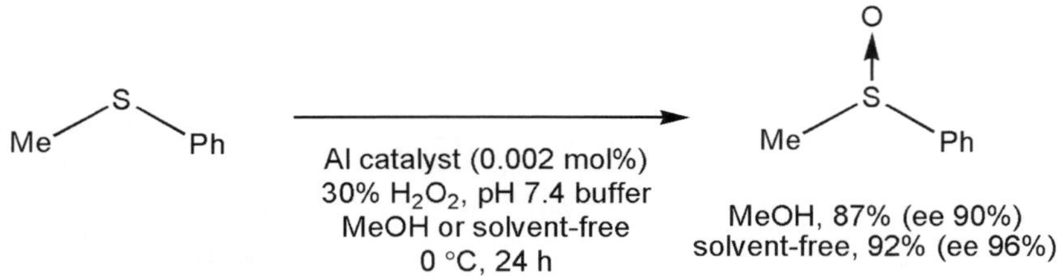

Figure 2.19 Asymmetric oxidation of sulfides under solvent free conditions.

Test reaction

Enantioselective aldol reaction

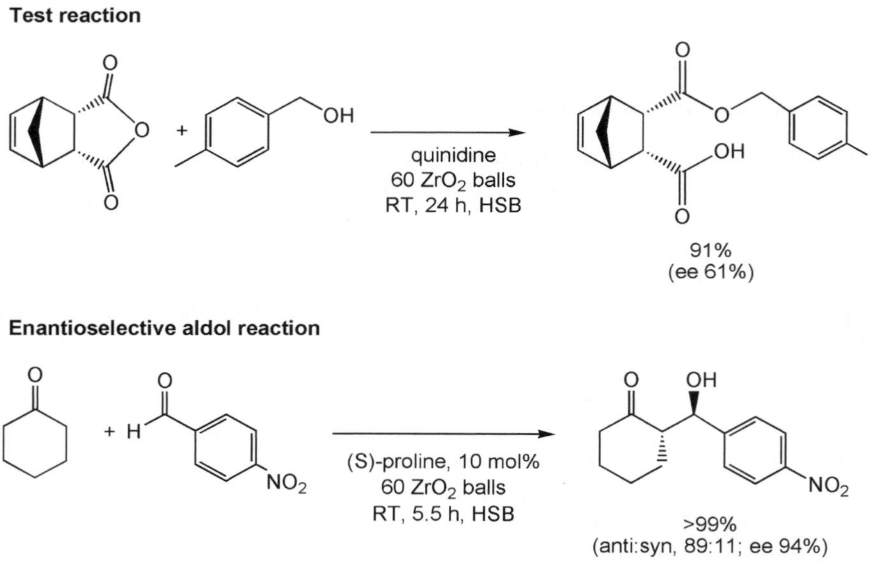

Figure 2.20 Enantioselective reactions using an HSBM.

higher under solvent free conditions than for analogous reactions performed in concentrated methanol solutions. However, it should be noted that aqueous hydrogen peroxide is used and therefore, although no added solvent is used, the reaction may have been aqueous in nature. Also, aqueous solutions, ethyl acetate and *n*-hexane were used in the product work up. Nevertheless, this is a green reaction and the very low catalyst loading is particularly noteworthy, as the enantioselective catalyst is the most expensive component of the chemical process.

A series of organocatalytic solvent free reactions have recently been performed using HSBM.[41,42] An asymmetric alkaloid-mediated opening of a cyclic *meso* anhydride was used as a test reaction (Figure 2.20). This reaction is normally performed at low temperatures in a VOC solvent. The results under solvent free conditions were comparable in terms of yield and enantioselectivity.

Having succeeded with this proof-of-concept reaction, Bolm and co-workers turned their attention to aldol reactions catalysed by (S)-proline. This reaction is conventionally performed in dimethyl sulfoxide (DMSO), a solvent that many green chemists are trying to replace. In the initial study, several ketones (cyclohexanone, cyclopentanone and acetone) were coupled with aldehydes containing both electron-withdrawing and electron-donating substituents.[42] The range of substrates investigated was expanded in a later study.[41] Generally, the reactions proceeded smoothly in good to excellent yields and afford the anti-aldol products with excellent diastereo- and enantioselectivities (Figure 2.20). Reactions were extracted from the vessel using diethyl ether or dichloromethane, and purified using pentane–ethyl acetate as eluents for flash chromatography. Therefore, although these results are very exciting, the greenness of the overall process could be improved by substituting diethyl ether, dichloromethane and pentane.

2.2.2.2   Microwave assisted Reactions

Microwave reactors have been used extensively in the field of solvent free syntheses (Figure 2.21) and this area has been the subject of two reviews by Varma.[6,7] In addition to requiring reduced work up because of the solvent free route, microwaves allow reaction times to be significantly reduced. The combination of these two areas leads to a 'win–win' situation for the organic chemist.

2.2.2.3   Photoreactions

Many photochemical reactions have also been performed under solvent free conditions.[2] These reactions are currently of less interest to typical synthetic chemists who are interested in producing molecules of significant biological or catalytic activity. However, photochemical reactions are likely to grow in importance during the coming decades, especially if sunlight can be used to aid the transformation. Additionally, many reactions that proceed easily through a photochemical route cannot be performed thermally. A selection of typical solvent free photochemical reactions is shown in Figure 2.22.

2.3   Summary and Outlook for the Future

Organic synthesis without solvents is already a mature field;[1] despite this, many chemists still assume that solvents are a necessity for most chemical processes. Therefore, the mindset of chemists needs to change and they must be willing to take up the opportunity that a solvent free method presents. Already, many multi-tonne industrial reactions are performed solvent free, particularly gas phase reactions such as ethylene polymerization. Although solid–solid reactions are yet to be performed on such a large scale, they have been performed on the kilogram scale.[5] Also, solvent free approaches have recently been introduced into the multi-step synthesis of a potential antituberculosis drug,

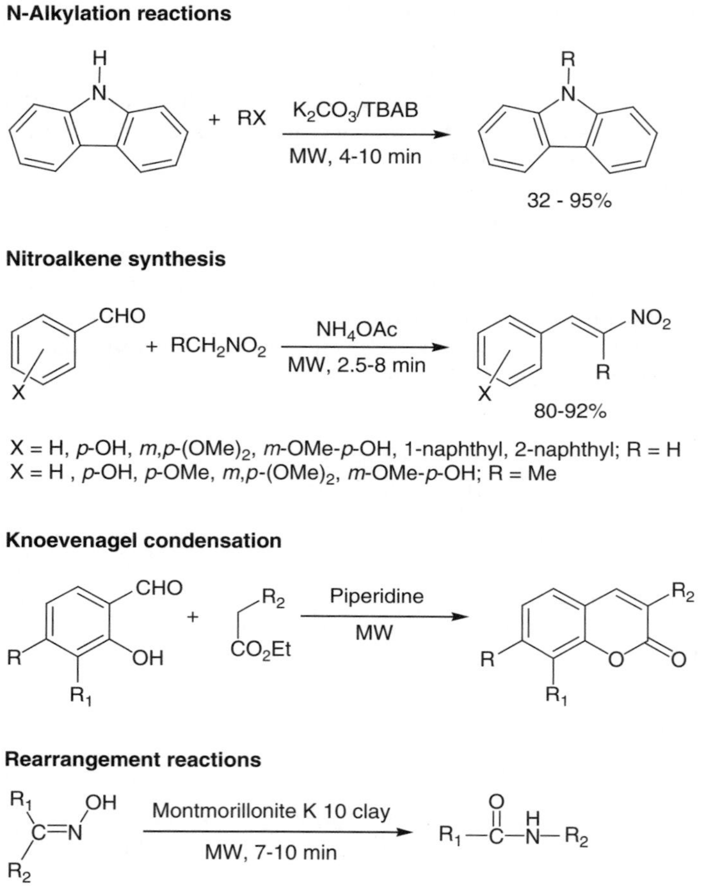
N-Alkylation reactions
+ RX
K_2CO_3/TBAB
MW, 4-10 min
32 - 95%
Nitroalkene synthesis
CHO
+ RCH_2NO_2
NH_4OAc
MW, 2.5-8 min
NO_2
80-92%
X = H, p-OH, m,p-$(OMe)_2$, m-OMe-p-OH, 1-naphthyl, 2-naphthyl; R = H
X = H , p-OH, p-OMe, m,p-$(OMe)_2$, m-OMe-p-OH; R = Me
Knoevenagel condensation
Piperidine
MW
CO_2Et
Rearrangement reactions
Montmorillonite K 10 clay
MW, 7-10 min

Figure 2.21 Some microwave assisted solvent free reactions.

PA-824.[43] The overall yield of the target compound was nearly tripled and the amount of solvent used was reduced by one third. A reduction in energy usage was also noted, as the extent of solvent removal between steps (in order to perform sequential reactions in different media) was significantly reduced. This study therefore demonstrates the great potential that solvent free reactions hold for complex organic procedures.

Another advantage of using no solvent (or less solvent), is that reaction times are often shorter, especially when a ball mill or microwave reactor is used. It is likely that solvent free methods will become more widespread as the number of microwave reactors and ball mills in research laboratories increases. For the green chemist, it is also worth noting that significant efforts need to be made in greening the work up of many of the reactions presented here and elsewhere. In most cases, any VOC solvent readily available is used, when a less hazardous or bio-sourced VOC would be a better option.

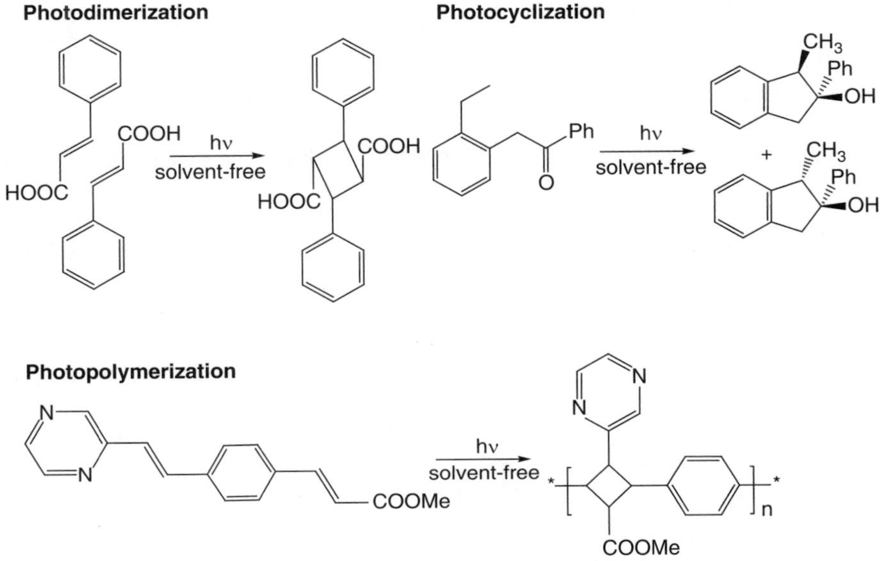

Photodimerization   **Photocyclization**

Photopolymerization

Figure 2.22 Some typical solvent free photochemical reactions.

In addition to organic syntheses, during the next 5 years solvent free methods will probably become more widely used in materials and inorganic chemistry as initial results in these areas are very promising. Also, although it may seem strange to think of analytical chemistry without solvents, as large volumes of eluent are used in chromatography, there is potential for their use in areas such as derivatization prior to gas chromatography analyses. Also, solvent free extraction methods should be further investigated. Juicing a plant reduces the volume of material for further separation and processing. On a large scale, this could potentially lead to a reduction in size of the processing site and reduced hazards. In summary, the biggest challenge to the advancement of solvent free methods is to change the minds of their potential users.

References

1. K. Tanaka, *Solvent-free Organic Synthesis*, Wiley-VCH, Weinheim, 2003.
2. K. Tanaka and F. Toda, *Chem. Rev.*, 2000, **100**, 1025.
3. T. Welton, *Green Chem.*, 2006, **8**, 13.
4. B. Rodriguez, A. Bruckmann, T. Rantanen and C. Bolm, *Adv. Synth. Catal.*, 2007, **349**, 2213.
5. G. Kaupp, *CrystEngComm*, 2006, **8**, 794.
6. V. Polshettiwar and R. S. Varma, *Acc. Chem. Res.*, 2008, **41**, 629.
7. R. S. Varma, *Green Chem.*, 1999, **1**, 43.
8. A. L. Garay, A. Pichon and S. L. James, *Chem. Soc. Rev.*, 2007, **36**, 846.

9. G. Kaupp, *CrystEngComm*, 2003, 117.
10. S. K. Li, Y. H. Shen, A. J. Xie, X. R. Yu, L. G. Qiu, L. Zhang and Q. F. Zhang, *Green Chem.*, 2007, **9**, 852.
11. S. Fernandes, C. S. R. Freire, C. P. Neto and A. Gandini, *Green Chem.*, 2008, **10**, 93.
12. K. Chen, C. Wang, D. Ma, W. Huang and X. Bao, *Chem. Commun.*, 2008, 2765.
13. P. J. Nichols, C. L. Raston and J. W. Steed, *Chem. Commun.*, 2001, 1062.
14. B. A. Roberts, G. W. V. Cave, C. L. Raston and J. L. Scott, *Green Chem.*, 2001, **3**, 280.
15. A. Orita, L. S. Jiang, T. Nakano, N. C. Ma and J. Otera, *Chem. Commun.*, 2002, 1362.
16. D. Braga, M. Curzi, A. Johansson, M. Polito, K. Rubini and F. Grepioni, *Angew. Chem. Int. Ed.*, 2006, **45**, 142.
17. D. Braga, M. Curzi, F. Grepioni and M. Polito, *Chem. Commun.*, 2005, 2915.
18. D. Braga, S. L. Giaffreda, F. Grepioni and M. Polito, *CrystEngComm*, 2004, **6**, 458.
19. G. M. Espallargas, L. Brammer, J. van de Streek, K. Shankland, A. J. Florence and H. Adams, *J. Am. Chem. Soc.*, 2006, **128**, 9584.
20. A. S. Filatov, A. Y. Rogachev and M. A. Petrukhina, *Cryst. Growth Des.*, 2006, **6**, 1479.
21. M. A. Petrukhina, A. S. Filatov, Y. Sevryugina, K. W. Andreini and S. Takamizawa, *Organometallics*, 2006, **25**, 2135.
22. A. Vaca, F. Teixidor, R. Kivekas, R. Sillanpaa and C. Vinas, *Dalton Trans.*, 2006, 4884.
23. B. C. Ranu, A. Hajra and S. S. Dey, *Org. Process Res. Dev.*, 2002, **6**, 817.
24. F. Fringuelli, R. Girotti, F. Pizzo, E. Zunino and L. Vaccaro, *Adv. Synth. Catal.*, 2006, **348**, 297.
25. J. Mack and M. Shumba, *Green Chem.*, 2007, **9**, 328.
26. E. Tullberg, F. Schacher, D. Peters and T. Frejd, *Synthesis*, 2006, **1183**.
27. E. Tullberg, D. Peters and T. Frejd, *J. Organomet. Chem.*, 2004, **689**, 3778.
28. G. W. Wang, K. Komatsu, Y. Murata and M. Shiro, *Nature*, 1997, **387**, 583.
29. T. H. Zhang, G. W. Wang, P. Lu, Y. J. Li, R. F. Peng, Y. C. Liu, Y. Murata and K. Komatsu, *Org. Biomol. Chem.*, 2004, **2**, 1698.
30. X. Moreau, J. P. Goddard, M. Bernard, G. Lemiere, J. M. Lopez-Romero, E. Mainetti, N. Marion, V. Mouries, S. Thorimbert, L. Fensterbank and M. Malacria, *Adv. Synth. Catal.*, 2008, **350**, 43.
31. B. Yan and Y. H. Liu, *Org. Lett.*, 2007, **9**, 4323.
32. M. M. Mojtahedi, E. Akbarzadeh, R. Sharifi and M. S. Abaee, *Org. Lett.*, 2007, **9**, 2791.
33. J. Mack, D. Fulmer, S. Stofel and N. Santos, *Green Chem.*, 2007, **9**, 1041.
34. N. Azizi, F. Aryanasab and M. R. Saidi, *Org. Lett.*, 2006, **8**, 5275.
35. K. M. Kuhn and R. H. Grubbs, *Org. Lett.*, 2008, **10**, 2075.
36. H. R. Hobbs and N. R. Thomas, *Chem. Rev.*, 2007, **107**, 2786.

37. P. J. Walsh, H. M. Li and C. A. de Parrodi, *Chem. Rev.*, 2007, **107**, 2503.
38. X. Zhang, H. Du, Z. Wang, Y. D. Wu and K. Ding, *J. Org. Chem.*, 2006, **71**, 2862.
39. K. Abdur-Rashid, A. J. Lough and R. H. Morris, *Organometallics*, 2001, **20**, 1047.
40. K. Matsumoto, T. Yamaguchi and T. Katsuki, *Chem. Commun.*, 2008, 1704.
41. B. Rodriguez, A. Bruckmann and C. Bolm, *Chem. Eur. J.*, 2007, **13**, 4710.
42. B. Rodriguez, T. Rantanen and C. Bolm, *Angew. Chem. Int. Ed.*, 2006, **45**, 6924.
43. A. Orita, K. Miwa, G. Uehara and J. Otera, *Adv. Synth. Catal.*, 2007, **349**, 2136.

CHAPTER 3
Water

3.1 Introduction

Beyond using no added solvent in a reaction or process, water is probably the greenest alternative we have. In fact, before the industrial and chemical revolutions in the nineteenth and early twentieth century, water was probably the most widely used medium for a wide range of applications. It is the most common molecule on the planet and therefore the cheapest solvent we can use, so it may seem somewhat surprising to non-chemists that it is not more widely used. To understand when and why water is an ideal solvent for some processes and when it would be detrimental, we must first consider its general properties as a solvent.

Some of the key physical properties of water are shown in Table 3.1. Data are also shown for acetone and ethanol, which are two relatively benign VOC solvents. On any of the chosen scales, whether it is dielectric constant or E^N_T, it is clear to see that water is a very polar solvent. It has a high dielectric constant, contains extensive hydrogen bonding and is a good Lewis base. This means that nearly all ionic compounds dissolve well in water by efficient solvation of the ions, and therefore any ion in water becomes associated with several water molecules. This solvation reduces the strength of attractive electrostatic forces between oppositely charged ions and allows them to separate and move more freely in solution. Solutions of salts are widely used as electrolytes and as buffer solutions. Deionized water has a pH of 7.0 but is amphoteric and partly dissociates in solution, giving rise to small amounts of H_3O^+ and OH^-. However, the pH is strongly affected by the presence of solutes with a high charge density, *e.g.* Al^{3+} forms acidic aqueous solutions. Although water is an excellent solvent for many inorganic species, it is also able to dissolve some organic molecules efficiently, *e.g.* sugars, proteins and low molecular weight acids.

Because of its extensive hydrogen bonding, the boiling point, melting point and critical points of water are much higher than those of acetone, ethanol and

RSC Green Chemistry Book Series
Alternative Solvents for Green Chemistry
By Francesca M. Kerton
© Francesca M. Kerton 2009
Published by the Royal Society of Chemistry, www.rsc.org

Table 3.1 Physical properties of water compared with acetone and ethanol.

	Water, H_2O	Acetone, $(CH_3)_2C=O$	Ethanol, CH_3CH_2OH
Melting point/°C	0	−94.7	−113.9
Boiling point/°C	100	56	78
Triple point/°C	0.01	−94.3	−123
Critical point, T_c/°C and P_c,/bar	374/221	235/48	241/63
Density at 25 °C/g cm^{-3}	1.00	0.78	0.78
Latent heat of vaporization/ kJ g^{-1} K^{-1}	2.26	0.518	0.846
Specific heat capacity/ J g^{-1} K^{-1}	4.19	2.44	2.15
Hildebrand solubility para-meter/(MPa)$^{0.5}$	47.9	19.7	26.2
Dielectric constant	78.30	20.7	24.3
Dipole moment	1.85	2.88	1.69
E^N_T	1.000 (defined)	0.654	0.355
α	1.17	0.08	0.86
β	0.47	0.43	0.75
π*	1.09	0.71	0.54
Donor number	1.46	0.44	0.82
Acceptor number	54.8	12.5	37.9

other organic solvents. The unusual structure of ice (solid water) further extends the special properties of this simple molecule, but is beyond the scope of this book.

There are many reasons why water is a desirable solvent and although some are hinted at in Table 3.1, a summary of these specific properties is provided in Table 3.2. However, there are also disadvantages when using water as a solvent, such as the low solubility of some compounds and the moisture-sensitive nature of many catalysts and reagents, which can lead to their deactivation. The high heat capacity of water, which is shown as an advantage in Table 3.2, can also be a disadvantage as it means that aqueous phases are difficult to heat or cool rapidly, and distilling water is very energy intensive. Also, although water and organic phases usually separate well, most organic compounds possess a small degree of solubility in water and this can lead to difficulties in purifying the aqueous phase after use. Therefore, care must be taken not to release con-taminated water into the environment.

The term *hydrophobic* is used to describe the apparent immiscibility of many organic compounds with water. However, the enthalpy of dissolution of a hydrocarbon in water is exothermic (ΔH is negative).[1] Therefore, the lack of miscibility must result from unfavourable changes in entropy. Because of the extensive hydrogen bonding in aqueous systems, water restructures itself around dissolved organic molecules to maximize hydrogen bonds. This ordering of the molecules leads to a decrease in entropy and a positive ΔG. If the organic compounds are grouped together and form a separate phase, the

Table 3.2 Summary of the advantageous properties of water as a solvent.

Safety advantages	Non-flammable
	Non-toxic
Reaction and process advantages	Low cost
	A density of $1\,\mathrm{g\,cm^{-3}}$ provides a sufficient difference from most organic substances for easy biphasic separation
	It is polar, so relatively easy to separate from apolar solvents or products; polarity may also influence (and improve) reactions
	Very high dielectric constant and therefore, favours ionic reactions
	High thermal conductivity, high specific heat capacity and high evaporation enthalpy
	High solubility of many gases, especially CO_2
	Highly dispersible and high tendency towards micelle formation in presence of suitable additives
Environmental advantages	Renewable
	Widely available in suitable quality (close to zero transportation costs)
	Odourless and colourless, making contamination easy to recognize

unfavourable entropy effect is reduced. The aggregation of organic molecules in reactions using water can lead to increases in reaction rates for some processes, and this will be discussed later in the chapter. The rate of some organic reactions is also influenced by the *salt effect*: organic compounds are less soluble in electrolyte solutions than in deionized water, as a result of an increase in the internal pressure caused by the presence of the salt.

3.1.1 Biphasic Systems

Using water as a reaction solvent can be an effective method of separating homogeneous catalysts from a reaction mixture and allowing them to be recycled and reused to give higher turnover numbers (TON) and reduce waste.[2,3] The low solubility of organic compounds in the aqueous phase can be overcome by using surfactants or phase transfer reagents. In the ideal process the organic substrate will be water soluble and the product insoluble, so separation will be easy. In addition to forming biphasic systems with many VOC solvents, alternative solvents such as fluorous media and supercritical carbon dioxide ($scCO_2$) can also be used and afford interesting biphasic systems.

In aqueous–organic biphasic catalysis, catalysts are used that will preferentially dissolve in the aqueous phase so that they can be recycled. The catalyst typically consists of a ligand and suitable metal salt. The ligands can be designed so that the resulting catalytic metal complex is hydrophilic, or at least water soluble (Figure 3.1). Additionally, vigorous stirring, a phase transfer

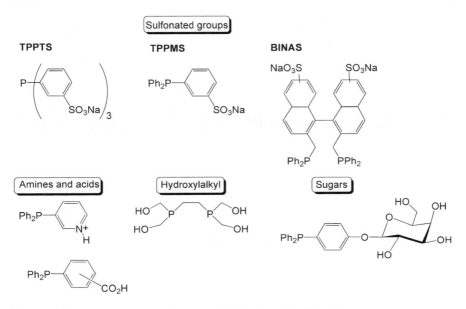

Figure 3.1 Some examples of water soluble phosphine ligands.

catalyst or a suitable surfactant can be used. However, not all catalyst systems are amenable to the aqueous biphasic approach. Many Lewis acid catalysts such as aluminium chloride are highly moisture sensitive and cannot be recycled and reused. Therefore, on an industrial scale they can lead to considerable waste and this has led to the development of new moisture-stable Lewis acid catalyst systems such as lanthanide triflates (the triflate anion is $CF_3SO_3^-$). In all aqueous–organic systems, it is important to remember that water is a potent nucleophile.

Sulfonated phosphines are perhaps the most widely used ligands in this field because they are soluble over a wide pH range, very poorly soluble in non-polar organic solvents, exhibit good stability and are easily prepared.[1] Also, phosphine ligands are common components in many transition metal catalysed reactions. Other classes of ligands, including amines, *N*-heterocyclic carbenes, tris(pyrazolyl)borates and porphyrins, have been rendered water soluble by adding suitable hydrophilic groups. Essentially, the presence of any group that can form strong hydrogen bonds is often sufficient to impart water solubility. Hydrophilic groups that have been used include hydroxyl, sugar, amine, acid and polyethylene glycol. Additionally, for *N*-containing ligands, such as the pyridinium-diphenylphosphine shown in Figure 3.1, the partitioning of the ligand and its metal complexes between the organic and aqueous phase can be adjusted by varying the pH.[1]

In the context of biphasic reaction systems, phase transfer catalysis should also be mentioned. It should be noted that it is not limited to aqueous–organic reactions or liquid–liquid systems, but is also sometimes employed in

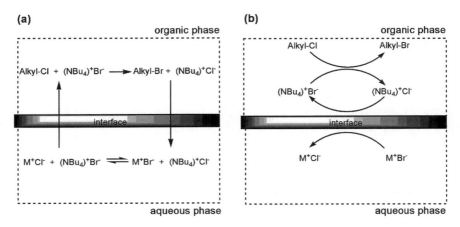

Figure 3.2   (a) A shuttling mechanism and (b) an interfacial mechanism for a simple
anion displacement reaction.

liquid–solid and liquid–gas reactions. A *phase transfer catalyst* (PTC) is defined
as a substance that will increase the rate of reaction between substrates in
separate phases.[4] There are two general types of PTC; organophilic cations
(quaternary ammonium and phosphonium salts) and polydentate complexing
agents (crown ethers, polyethylene glycols and cryptands). In the context of
green chemistry, it should be noted that polyethylene glycols are less expensive
and less toxic than crown ethers and cryptands. However, a simple and colour-
ful demonstration of the principle of phase transfer can be performed using
18-crown-6, aqueous potassium permanganate and a non-polar organic solvent
(historically, benzene). The crown ether dissolves in the organic solvent, the
potassium ion complexes with the crown ether, and the permanganate is forced
to enter the organic phase (leading to a purple coloration) in order to ion-pair
with the potassium ion. Although this is a useful demonstration, it is also
possible that PTCs act through an interfacial mechanism where the organophilic part of the PTC (*e.g.* ammonium cation) never enters the aqueous phase.
Both mechanisms are shown schematically in Figure 3.2 and the mechanism
employed will vary on a case-by-case basis depending on the nature of the PTC,
the organic solvent and reagents.

   All PTCs have a number of advantages and disadvantages (Table 3.3), and
often the system will have to be optimized for a particular reaction to avoid
emulsion formation and other problems. However, in most cases, the increase
in reaction rate and decrease in the amount of solvent used are significant and
often outweigh the disadvantages. Typical reactions that work well under PTC
conditions include simple nucleophilic substitutions, Friedel–Crafts reactions,
Wittig reactions and oxidations.

   In order to overcome toxicity issues with some PTCs, new bio-sourced and
biocompatible surfactants are being developed.[5]

Table 3.3 Summary of advantages and disadvantages of phase transfer catalysis.[1]

Advantages	Disadvantages
Reduced need for organic solvents	Catalyst required, which may be toxic
Relatively inexpensive auxiliaries required, *e.g.* $(NBu_4)^+Cl^-$	Difficult to separate; emulsions may form
Improved separation possible	Catalyst can be hard to recover
Increased reaction rates and productivity	Vigorous mixing required
Improved selectivity, due to lower operating temperatures	Contaminated waste water can be tough to purify

3.2 Chemical Examples

3.2.1 Extraction

Many a chemistry student, past or present, will remember using water when performing two different extraction methods. First, steam distillations are used to isolate thermally sensitive compounds including some natural products. Second, separatory/separating funnel techniques including simple aqueous–organic (liquid–liquid) separations form the basis for many of the facile separations (including catalyst recycling studies) that are described for organic processes later in this chapter. More complicated variants of aqueous–organic separations are also commonplace such as those involving sequential acid-ification and basification for purification of molecules, including alkaloids, that contain amine and other basic functional groups. Needless to say, water is used extensively in this way on a large scale as well as in academic laboratories.

Steam distillation is used when normal distillation is not an option, due to thermal sensitivities. By adding water or steam, the boiling points of compounds can be depressed, allowing them to evaporate at lower temperatures. After distillation the vapours are condensed and typically yield two phases: water and organics. These can then be easily separated on the basis of their different densities. On a simple level, this can be seen every day in fresh black coffee, where small amounts of organic, water-immiscible oils float on the surface. Steam distillation is employed on an industrial scale in the manufacture of essential oils (*e.g.* lavender, eucalyptus and fruit oils) that are used in the food and flavour industries. In this area, the use of water is highly desirable as a natural solvent. Steam distillation is also used in petroleum refineries and petrochemical plants, where it is commonly referred to as 'steam stripping'.

Recent advances in the area of steam distillation have come with the intro-duction of microwave heating,[6,7] which further reduces the decompositional change of thermally unstable compounds. Microwave assisted extraction of lavender essential oil, using the apparatus shown in Figure 3.3, provided advantages over traditional steam distillation in terms of saving energy, redu-cing time (10 min *vs* 90 min), product yield, cleanliness and quality.[7] A com-parative study of greener extraction techniques also reported on the use of microwave heating in the extraction of the essential oil from *Artemisia afra*.[6]

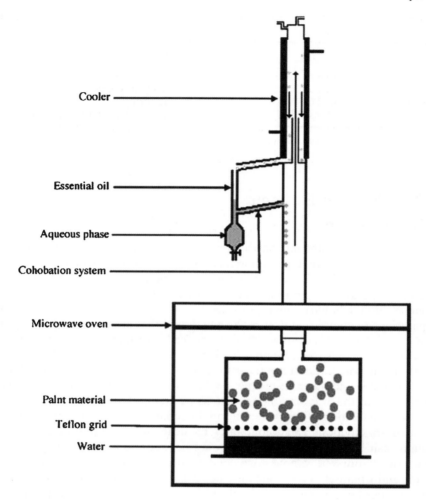

Figure 3.3 Schematic diagram of the apparatus used for microwave accelerated steam distillation (MASD). [Reprinted with permission from *Anal. Chim. Acta,* 2006, **555**, 157–160. Copyright 2006 Elsevier B.V.]

This plant (found in east, central and southern Africa) is traditionally used for its fragrance, as an insect repellent and as a medicinal herb. The amount of oil obtained after 10 min using microwave heating was comparable to that obtained after 3 h of traditional distillation. The composition of the oil was found to be comparable to that found in the distillations of lavender oil. However, in this study, the concentration of yogomi oil yielded through a traditional steam distillation was significantly higher than that from microwave assisted extraction (Figure 3.4). This could be due to increased thermal rearrangements caused by the prolonged heating required for traditional steam distillation.

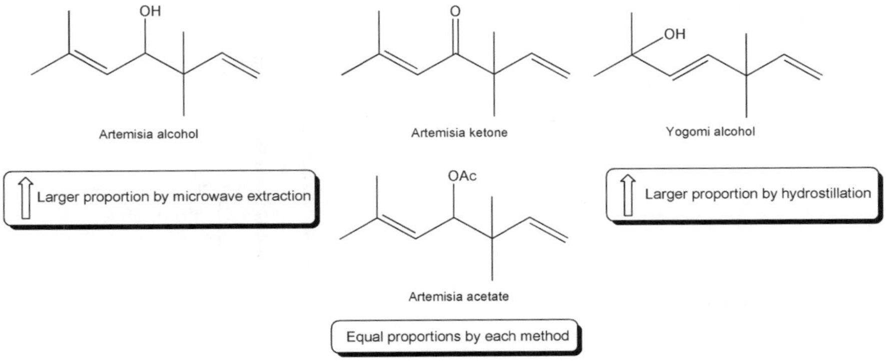

Figure 3.4 Oxygenated components of essential oils extracted from *Artemisia afra* by hydrodistillation and microwave extraction.

3.2.2 Chemical Synthesis

Of all the alternative reaction media, water has perhaps been used most extensively as modifications to reagents are rarely required, and often the organic substrates do not even need to be soluble in the aqueous phase for the reaction to proceed smoothly. Organic synthesis in water has been extensively reviewed during the last 5 years,[8–14] therefore just an overview of the field will be given here and interested readers should refer to the review articles for more information on this rapidly growing field. Water has proven itself as a very useful solvent for many types of reactions, including Diels–Alder, aldol, other carbon–carbon bond-forming reactions including C–H activation processes, epoxidation reactions and alcohol oxidations (Figure 3.5). However, organic chemistry in water is by no means perfect or simple. Problems can arise as a result of the low solubility of organic substrates and catalysts. Also, difficulties can arise in removing trace amounts of organic impurities from the aqueous phase. This might potentially be overcome by using a downstream superheated or supercritical water oxidation process to remove the organic contaminants.

Breslow and co-workers have performed some of the most outstanding work in this field and their results initiated a flurry of research. They found that the rates of reaction and selectivity in the Diels–Alder reactions are improved in an aqueous system (Figure 3.6).[15] Additionally, the presence of salts or β-cyclodextrins can enhance the hydrophobic effect, which causes organic molecules to cluster together in aqueous solution, and further accelerate the Diels–Alder reaction. It should also be noted that related photochemical [2 + 2] additions can be performed using water, and in some cases these show similar rate enhancements due to hydrophobic effects.[12]

More recently, the hydrophobic effect has been used in its most extreme form, where the organic substrates are hydrophobic and insoluble; such reactions are said to proceed 'on water'.[16–19] As can be seen in Figure 3.6, yields are significantly improved compared with the same reactions in VOC solvents and

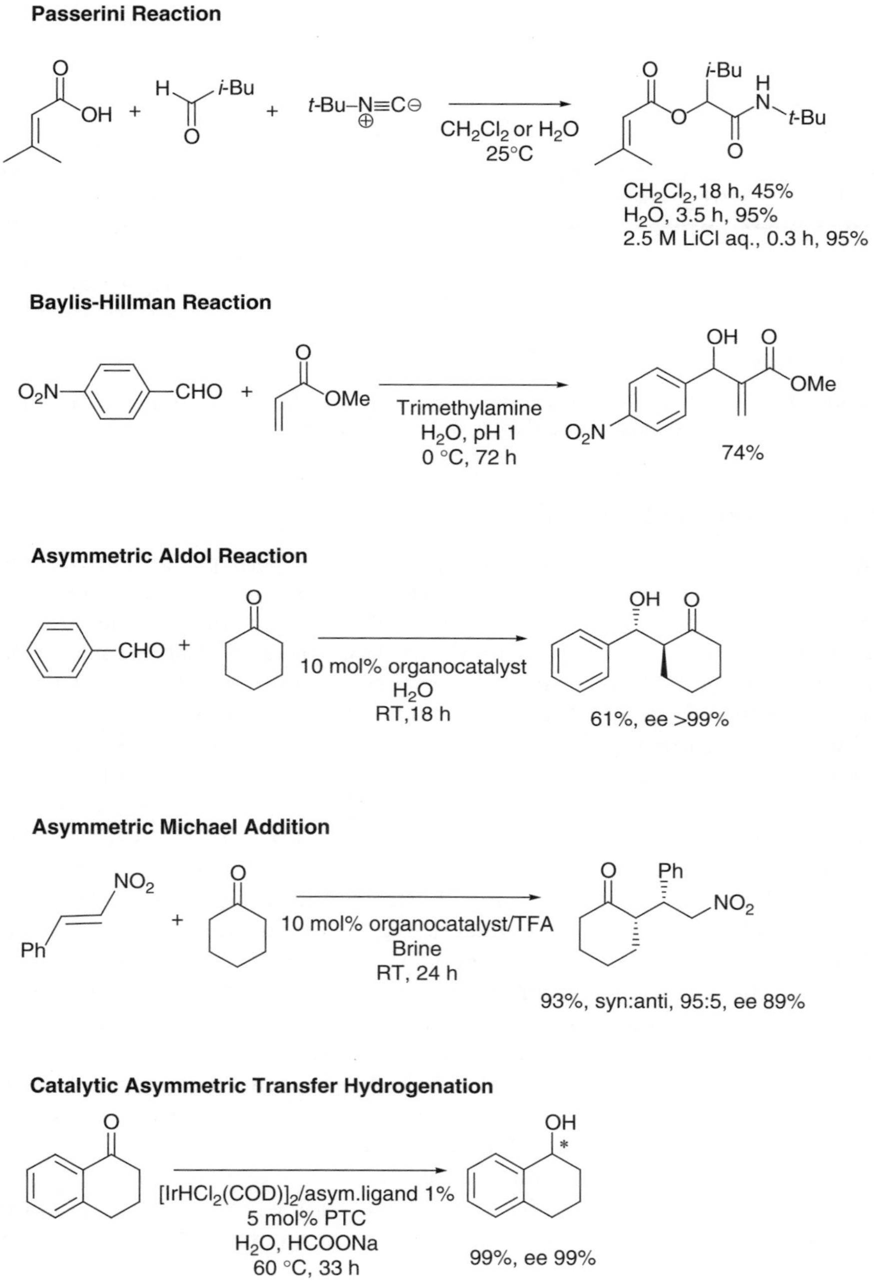
Passerini Reaction
OH
H
i-Bu
t-Bu–N≡C⊖
⊕
CH2Cl2 or H2O
25°C
i-Bu
H
N
t-Bu
CH2Cl2,18 h, 45%
H2O, 3.5 h, 95%
2.5 M LiCl aq., 0.3 h, 95%
Baylis-Hillman Reaction
O2N
CHO
OMe
Trimethylamine
H2O, pH 1
0 °C, 72 h
OH
OMe
O2N
74%
Asymmetric Aldol Reaction
CHO
10 mol% organocatalyst
H2O
RT,18 h
OH
61%, ee >99%
Asymmetric Michael Addition
NO2
Ph
10 mol% organocatalyst/TFA
Brine
RT, 24 h
Ph
NO2
93%, syn:anti, 95:5, ee 89%
Catalytic Asymmetric Transfer Hydrogenation
[IrHCl2(COD)]2/asym.ligand 1%
5 mol% PTC
H2O, HCOONa
60 °C, 33 h
OH
99%, ee 99%

Figure 3.5 Some synthetic organic reactions conducted in aqueous media.

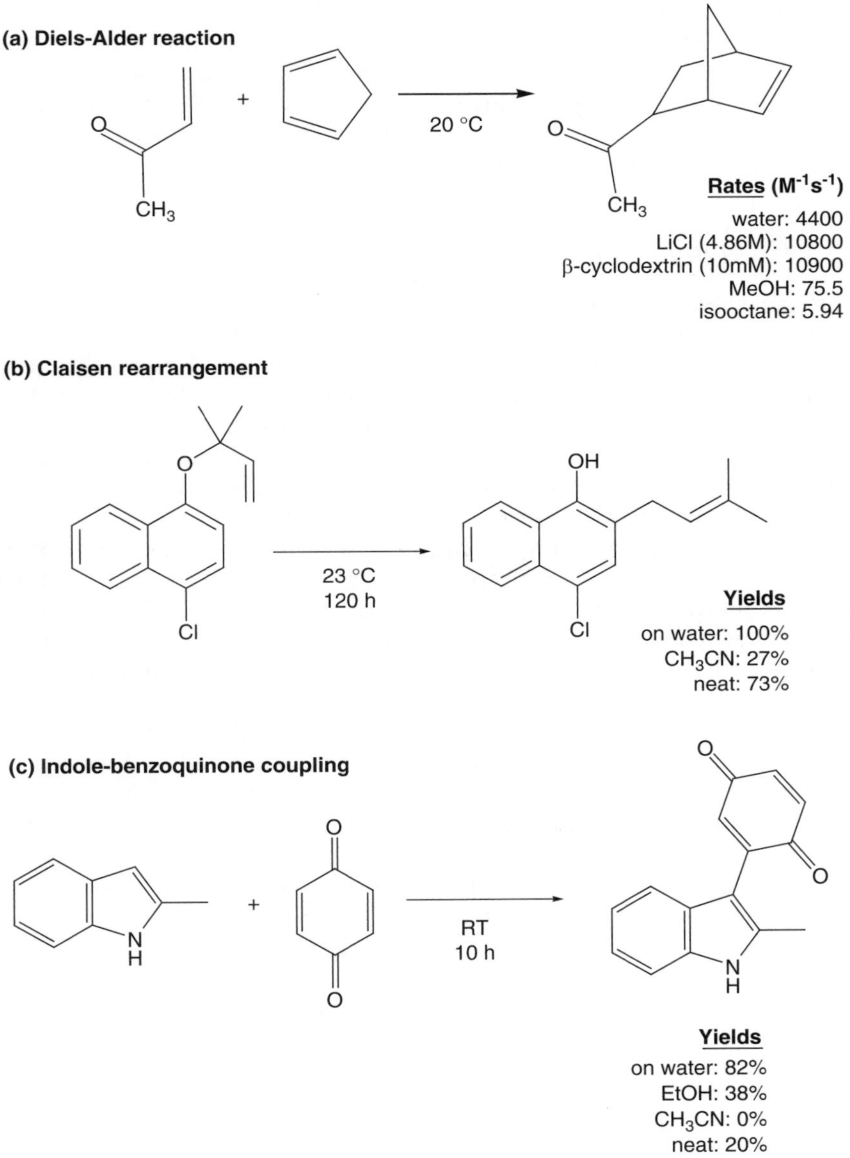

(a) Diels-Alder reaction

20 °C

Rates (M⁻¹s⁻¹)

Rates (M^{-1}s^{-1})
water: 4400
LiCl (4.86M): 10800
β-cyclodextrin (10mM): 10900
MeOH: 75.5
isooctane: 5.94

(b) Claisen rearrangement

23 °C
120 h

Yields
on water: 100%
CH$_3$CN: 27%
neat: 73%

(c) Indole-benzoquinone coupling

RT
10 h

Yields
on water: 82%
EtOH: 38%
CH$_3$CN: 0%
neat: 20%

Figure 3.6 (a) An example of a Diels-Alder reaction accelerated by a hydrophobic effect; (b) a rearrangement and (c) a coupling reaction 'on-water' performed.

are also greater than those achieved under solvent free conditions. In 2008, Cozzi and Zoli reported on the nucleophilic substitution of alcohols 'on water'; no surfactants or acid catalysts were required.[19] Around the same time, Shapiro and Vigalok reported aerobic oxidation of aldehydes 'on water'.[18]

These proceeded in good to excellent yields just by stirring the emulsion in air; no catalyst or other additive was required.[18] The organic acid obtained could then be reacted in the same pot 'in water' in a Passerini reaction with an iso-cyanide and some unoxidized aldehyde. This study exemplifies an interesting control mechanism for tandem reactions, by using the phase behaviour of the reagents to control the reactivity.

3.2.2.1   Metal-mediated and Catalysed Reactions

After polymerization processes, one of the most important aqueous phase reactions to be performed on an industrial scale is the Rhone-Poulenc hydro-formylation process that utilizes a water soluble rhodium phosphine catalyst.[2] This process will be discussed in more detail in Chapter 10. The success of this process has led to many exciting results in metal catalysed aqueous phase chemistry. Additionally, amazing advances have been made where reactions that are typically considered unsuited to the presence of moisture, *e.g.* Grignard-type chemistry, can be performed in water.

It should be noted that in most cases reactions are performed in deionized water. However, this is not always necessary, as evidenced by the efficient catalytic syntheses of quinoxalines in tap water (Figure 3.7).[20] The products from the condensation reaction can be isolated in 98% purity by simple fil-tration and analytically pure products can be obtained through dissolution in ethyl acetate and passage through a small plug of silica.

Lewis acids are widely used in organic catalysis. Unfortunately, many tra-ditional Lewis acids are highly moisture sensitive, which led to investigations for stable alternatives. It has been found that salts of rare earth metals (*e.g.* ytterbium triflate) are one such group of alternatives. It was recently discovered that rate enhancements could be achieved by adding small amounts of ligand to ytterbium triflate catalysed Michael addition reactions in water.[21] Although in this study the recycling of the catalyst was not studied, it is likely that these species can be easily recycled in a similar way to related nitration catalysts previously reported.[22] More recently, this work has been extended and it has been shown that an α-amino acid induced rate acceleration in aqueous biphasic Lewis acid catalysed Michael addition reactions (Figure 3.8).[23] The high water

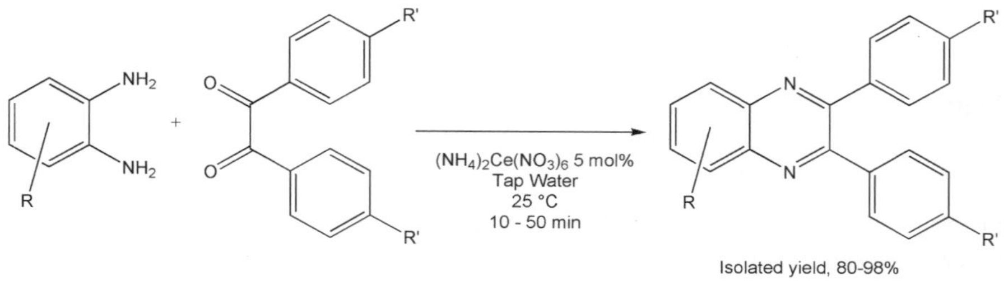

Isolated yield, 80-98%

Figure 3.7   Synthesis of quinoxalines in tap water.

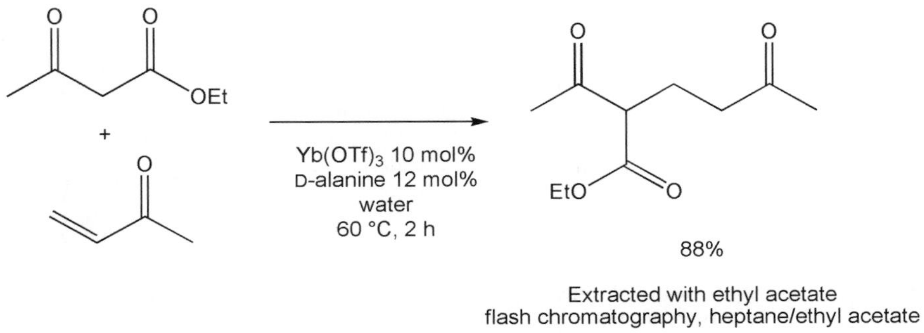

Figure 3.8 Lewis acid catalysed Michael addition.

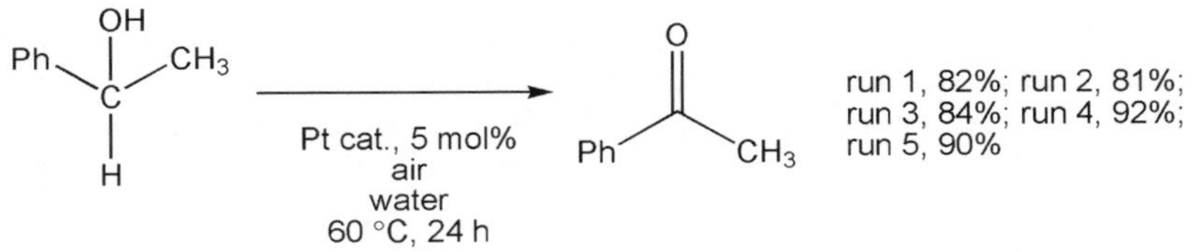

Figure 3.9 Aerobic oxidation of alcohols using a nanoplatinum catalyst.

solubility of the two-component ytterbium triflate–alanine catalyst allowed easy isolation of the Michael adducts and the aqueous catalyst phase could be recycled four times with no loss in activity. In addition to using water as a solvent, this reaction is of interest to green chemists as it uses a natural α-amino acid as a chiral ligand. Needless to say, further studies in this area are ongoing.

Phase transfer catalysis can also be used to enhance a metal catalysed process in the aqueous phase if insolubility of the catalyst is an issue. For example, kinetic resolution of secondary alcohols can be achieved using chiral manganese oxidation catalysts that are insoluble in water in combination with tetraethylammonium bromide.[24] In the absence of the PTC enantioselectivity was negligible, but it increased to 88% when the PTC was used. A water soluble diruthenium complex has also been used as a recyclable catalyst in the aerobic oxidation of a range of primary and secondary alcohols.[25] The catalyst-containing aqueous phase was reused three times without any loss in activity. Also, as the rate of reaction was 14 times greater for the oxidation of primary alcohols than for the oxidation of secondary alcohols, there is the potential of selectively converting primary alcohols to aldehydes in the presence of secondary and tertiary alcohols. More recently, supported platinum nanoparticle catalysts have been developed for the aerobic oxidation of a wide range of alcohols in aqueous solution (Figure 3.9).[26] The particles were embedded within an amphiphilic resin (polystyrene–polyethylene glycol), were highly reactive at a relatively low temperature and were efficiently recycled four times. The organic product

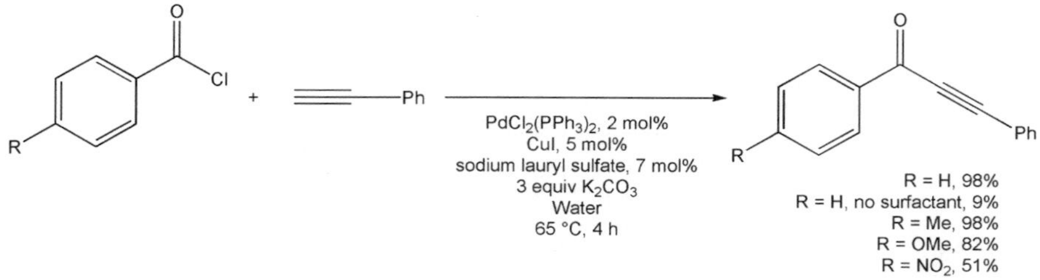

Figure 3.10   Synthesis of ynones through the coupling of terminal alkynes and acid chlorides in water.

generally partitions into the organic polymer bead from the aqueous solution and therefore, in order to maintain 'greenness', supercritical carbon dioxide was used to remove the product.

Some of the most widely studied organic reactions at this time are palladium catalysed carbon–carbon cross coupling reactions, which have been extensively investigated in water. For example, palladium catalysed Suzuki reactions can be performed in water in the presence of poly(ethylene glycol) (PEG).[27] It should be noted that the PEG may be playing the role of a surfactant (PTC) and/or a support for the metal catalyst in water. Interestingly, in this example, no phosphine is needed and the products are easily separated and the catalyst phase reused. Unfortunately, diethyl ether was used to extract the product and as this solvent is hazardous (low flash point and potential peroxide formation), the overall process would be greener if an alternative solvent could be used.

Another palladium catalysed reaction that has been successfully performed in water is the direct coupling of acid chlorides with alkynes.[28] Copper is used as a co catalyst and the choice and use of a surfactant are essential to the success of the reaction (Figure 3.10).

3.2.2.2   Microwave Assisted Reactions

The use of microwave irradiation as a heating source in combination with water as a solvent was recently reviewed.[9] Unfortunately, in many cases comparative studies using conventional heating or VOC solvents are not reported. However, despite this lack of data, it is clear that there are special benefits (particularly regarding time) to performing reactions under these conditions. For example, in challenging transition metal catalysed coupling reactions time can be reduced from hours or days to minutes, and if the reaction is performed in a sealed vessel there is often no need to apply an inert atmosphere. Reactions studied to date utilizing both water and microwave heating include carbon–carbon couplings (Suzuki, Heck, Sonogashira, *etc.*), carbonylations, hydrogenations, heterocycle syntheses, Mannich-type reactions, nucleophilic substitutions, ring-openings of epoxides and many more (Figure 3.11). Particularly noteworthy are phosphine-free, low palladium loading, carbon–carbon coupling reactions that have been developed by the Leadbeater group.[29]

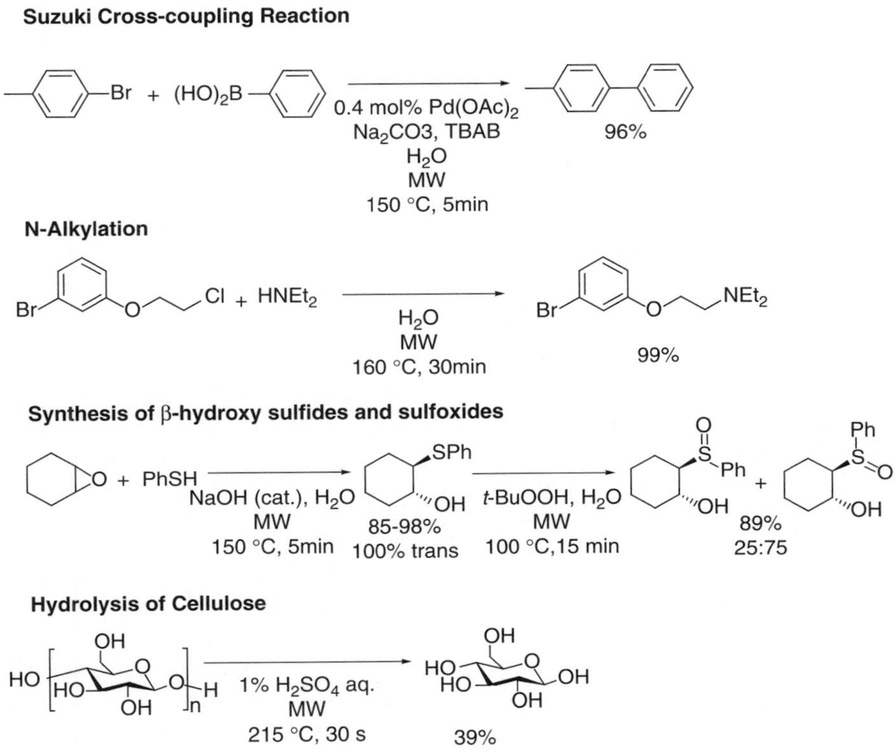

Suzuki Cross-coupling Reaction

N-Alkylation

Synthesis of β-hydroxy sulfides and sulfoxides

Hydrolysis of Cellulose

Figure 3.11 Some microwave assisted reactions using water as the solvent.

In addition to organic reactions, acid catalysed hydrolysis of cellulose has been performed in a rapid and controlled manner using a microwave reactor.[30] Given this reaction, it is likely that aqueous phase microwave assisted reactions will play an important role in the rapid development of biorefinery based materials and chemicals.

3.2.2.3 Biocatalysis

Enzymes as nature's catalysts are able to perform an outstanding array of regio- and stereoselective reactions. Therefore, as water is nature's solvent, it is not surprising that many biocatalytic reactions have been performed in the aqueous phase.[31] However, in typical reactions, the substrates are limited to hydrophilic compounds because of a desire for reaction homogeneity. It should also be noted that, in most cases, the aqueous medium is a buffer solution of an ideal pH for the enzyme to function effectively.

Reaction studies include cyanations using hydroxy nitrile lyases, hydrolysis of amides using acylases, amidases or lipases, and even dehydration reactions of aldoximes to nitriles using aldoxime dehydratase. This reaction is quite

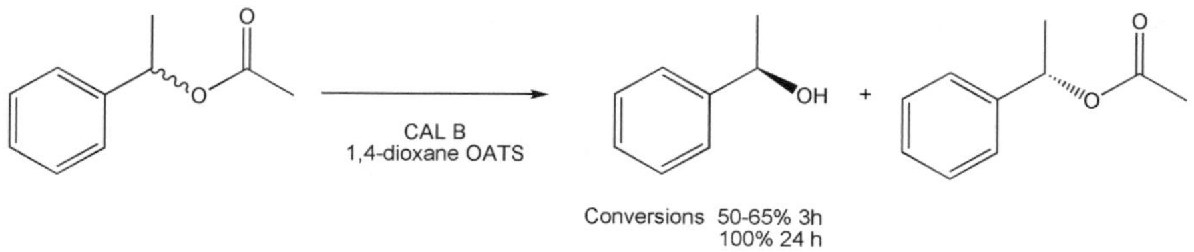

Conversions 50-65% 3h
100% 24 h

Figure 3.12   Biocatalysis using organic–aqueous tunable solvents (OATS).

remarkable given the large excess of water and tendency for biocatalytic hydrolysis reactions in this medium. Carbon–carbon bond-forming reactions can also be performed by enzymes in aqueous media including carboxylations using decarboxylase enzymes and aldol reactions using aldolases. Importantly, because of the excellent substrate selectivity of enzymes, dynamic kinetic resolutions can be performed where only one enantiomer of a racemic mixture will be converted to yield the product in an enantiomerically pure form. This excellent selectivity is clearly the largest driving force in the development of new biocatalytic processes and the isolation and evolution of new enzymes.

A recent advance in this area is the development of *organic aqueous tunable solvents* (OATS) for biocatalytic reactions and catalyst recycling.[32,33] This allows hydrophobic substrates to be transformed by using a small portion of water-miscible organic solvent in the reaction mixture. Upon completion of the reaction, the mixture is exposed to a carbon dioxide pressure of 10–50 bar, which induces phase separation. The products enter the organic phase and can be separated from the aqueous catalyst-containing phase, thus allowing facile recycling. The process has been successfully used in hydrolysis reactions catalysed by *Candida antarctica* lipase B (CAL B) and the kinetic resolution of rac-1-phenylethyl acetate to (*R*)-1-phenylethanol (Figure 3.12). Further information on different tunable solvent systems is given in Chapter 9.

3.2.2.4   *Carbon Dioxide Fixation*

Using carbon dioxide as a feedstock in synthetic chemistry is an important area of green chemistry. It is significantly soluble in water, and water is therefore a good medium for its conversion. However, when it dissolves it forms carbonic acid (Figure 3.13). Considerable efforts have been made to understand this process and control the pH of aqueous–carbon dioxide systems.[34] This is also highly relevant to studies involving supercritical carbon dioxide and water in biphasic catalysis, especially for pH-sensitive enzymes.

Eghbali and Li have recently reported a highly efficient method for the conversion of carbon dioxide to cyclic carbonates in water (Figure 3.14).[35] The organic base 1,8-diazabicyclo[5.4.0]undec-7-ene (DBU) was used as a simple 'carbon dioxide activator' at a level of 0.1–0.3 equivalents per mole of alkene. The reaction is catalysed by a catalytic amount of bromine provided by tetra-butylammonium bromide (TBAB) or sodium bromide. During the course of

$$CO_2 (g) \rightleftharpoons CO_2 + H_2O \qquad (1)$$

$$CO_2 + H_2O \rightleftharpoons H_2CO_3 \qquad (2)$$

$$H_2CO_3 \rightleftharpoons H^+ + HCO_3^- \qquad (3)$$

$$H^+ + HCO_3^- \rightleftharpoons 2H^+ + CO_3^{2-} \qquad (4)$$

$$[H^+] = \frac{K_1[H_2CO_3]}{a(HCO_3^-)}$$

Figure 3.13 Carbon dioxide–water–carbonic acid equilibria present in aqueous–carbon dioxide systems.

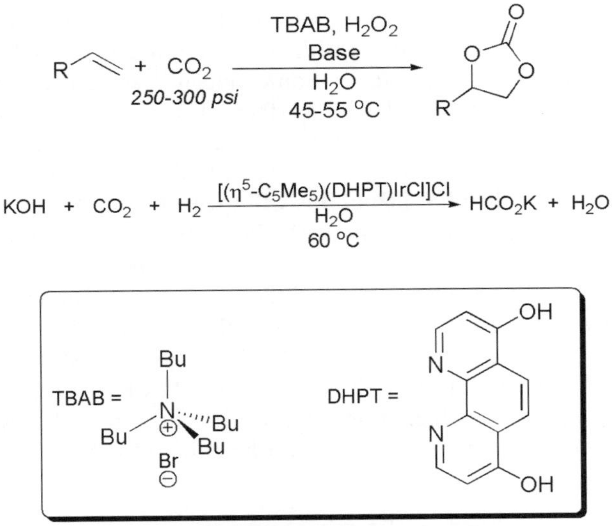

Figure 3.14 Carbon dioxide conversions in aqueous media.

the reaction, hydrogen bromide is formed and re-oxidized by aqueous hydrogen peroxide to continue the catalytic cycle. In this study, the organic product was extracted using ethyl acetate and purified by column chromatography. However, there is the opportunity for further optimization and this metal-free catalytic conversion deserves further investigation.

A range of metal catalysts have also been studied in aqueous solution for the transformation of carbon dioxide, including rhodium, ruthenium and iridium bipyridine or phenanthroline complexes.[36] One of the most effective systems is the iridium complex shown in Figure 3.14. The ligand design concept used in this study is very clever. The catalytic activity of the complex and its solubility in aqueous solution can be tuned by the pH of the solution.[37] Under acidic conditions, DHPT is protonated (pyridinol form) but under basic conditions, it exists as an oxyanion (pyridinolate form). The ligands containing hydroxy substituents yield complexes that are up to 100 times more active in this reaction than conventional bipyridine and phenanthroline analogues. The change in solubility with pH has allowed these catalysts to be recycled efficiently four times with low levels of iridium leaching (0.1–0.6 ppm).

These studies show that there is significant scope for activation and fixation of carbon dioxide in the aqueous phase.

3.2.3 Materials Synthesis

Water has been widely used as a solvent for polymer and materials synthesis. In fact, aqueous media are used for over 50% of radical polymerizations industrially.[38] It is estimated that 10 million tonnes of polymer are produced this way annually and are primarily used in environmentally friendly coatings and paints. The polymerizations are generally conducted as oil-in-water emulsions, where the monomer is suspended inside a micelle and undergoes polymerization within this organo-rich phase. Some micellar structures are shown in Figure 3.15. Micelles are formed when surfactants (anionic, cationic or neutral) are dissolved at a certain concentration (the critical micelle concentration) or higher. The surfactants consist of two parts; a hydrophilic and a hydrophobic part. In water, the hydrophobic sections will form the interior of the micelle to minimize unfavourable interactions with the water and maximize favourable interactions between the water and the hydrophilic head. The monomer will enter the interior, organophilic, part of the micelle to minimize hydrophobic interactions and the growing polymer particle will also be stabilized there. These emulsion polymerizations lead to the formation of a stable dispersion of polymer particles in water (known as a latex), and they have been conducted for a wide range of monomers including styrene, vinyl acetate and acrylic acid. The size and shape of the polymer particle can be controlled by the concentration

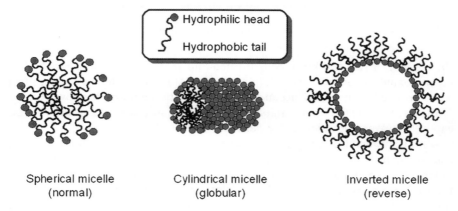

	● Hydrophilic head	
	⌇ Hydrophobic tail	

Spherical micelle   Cylindrical micelle   Inverted micelle
(normal)   (globular)   (reverse)

Figure 3.15 Some micellar structures that can form in aqueous–organic systems.

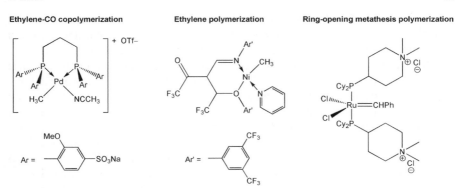

Figure 3.16 Some transition metal catalysts for aqueous polymerizations.

and type of surfactant used. If the water is a minor component in a surfactant–organic solution, inverted micelles can form where the hydrophobic groups form the corona (or exterior) of the micelle. The formation of micelles (normal and reverse) also plays an important role when using supercritical fluid reaction media (Chapter 4).

The development of atom-transfer radical polymerization processes[39] that can be conducted in aqueous media has had a significant impact on this field. Copper catalysts with organic initiators can perform living polymerizations of a large array of monomers, and block copolymers and other functional structures can be prepared in this way. More recently the types of polymers which can be prepared in water have further increased as water soluble and -stable metal catalysts are developed.[38] Reactions studied include copolymerizations of carbon monoxide and olefins, ethylene and α-olefin homopolymerizations and ring-opening metathesis polymerizations (Figure 3.16). These can be conducted as precipitation polymerizations (where the monomer is sufficiently soluble in water and the polymer is insoluble), homogeneous polymerizations (where the monomer and polymer are both water soluble) or in emulsions. The nickel complex shown in Figure 3.16 was recently used in the emulsion polymerization of ethylene using sodium dodecylsulfate (SDS) as the surfactant.[40] Its average activity was 1.9×10^4 turnovers per hour and polymer molecular weights of 30 000–50 000 were obtained.

As a result of the advances in catalyst discovery for aqueous ethylene polymerization, silica–polyethylene nancomposites have been prepared with structures that vary with changing catalyst structure and silica composition.[41] It is likely that many more advances in the area of high-tech composites with potential biological and nanotechnology applications will be made in the next few years through aqueous polymerization processes. In addition to free radical polymerizations and catalytic polymerizations, it should be noted that oxidative polymerizations can also be performed in aqueous media to yield conducting polymers. Recently, this has been used to prepare polypyrrole-coated latex particles that are expected to be interesting synthetic mimics for micrometeorites.[42]

Another area of materials chemistry in which aqueous phase transformations can play an important role is the formation of metallic nanoparticles through reduction of metal salts in the presence of suitable stabilizing agents. This can involve traditional surfactants as used in emulsion polymerization to stabilize particle growth, such as Triton X-100 (*p-tert*-octylphenoxy polyethylene).[43] Carbon dioxide has been used recently to separate this surfactant from gold nanoparticles prepared in aqueous solution; this will be discussed in more detail in Chapter 4.[43] However, in other recent studies researchers have looked into greener synthetic methods of synthesis and the use of alternative natural reducing agents instead of potentially hazardous sodium borohydride. In 2003, Wallen and co-workers reported a method of preparation for starched silver nanoparticles.[44] Aqueous silver nitrate is added to a solution of soluble starch and then the reducing sugar β-D-glucose is added. Upon heating and stirring at 40 °C for 20 h, the solution becomes yellow in colour and silver particles of an average diameter of 5.3 nm are formed. In this example, the starch is acting as a template and its surface hydroxyl groups act to stabilize the particles. This approach was subsequently extended to the synthesis of starched gold nanoparticles, through reduction of chlorauric acid ($HAuCl_4 \cdot 3H_2O$), and silver–gold alloy nanoparticles.[45] It has subsequently been shown by Ikushima and co-workers that gold nanoparticles can be obtained in the absence of stabilizing starch and glucose can form the stabilizing layer or coating.[46] This has been inferred through IR analysis of the resulting nanoparticles. Additionally, they also demonstrated catalysis using the resultant gold nanocrystals (Figure 3.17). In the absence of gold nanoparticles, the 4-nitrophenol could not be reduced by sodium borohydride.

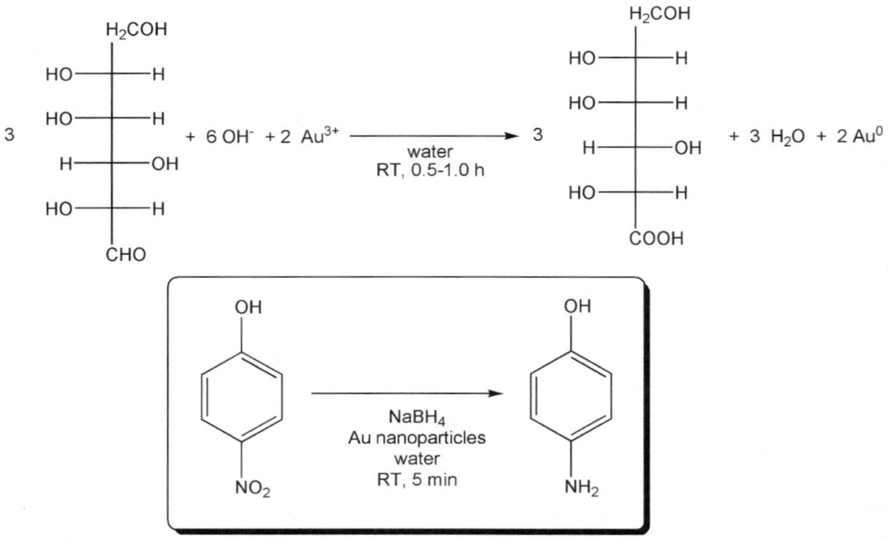

Figure 3.17 Preparation of gold nanoparticles in aqueous solution using glucose as the reducing agent and their catalytic activity in reduction of *p*-nitrophenol.

Silver nanoparticles have also been prepared in aqueous solution using *Capsicum annum* L. extract.[47] It is thought in this example that Ag(I) is reduced to Ag(0) by proteins within the natural extract and that these proteins also act to stabilize the particles. The size of the nanoparticles was found to increase with reaction time: 5 h, 10 ± 2 nm; 9 h, 25 ± 3 nm; 13 h, 40 ± 5 nm. It should be noted that gold and silver nanoparticles have potential pharmaceutical and biomedical applications, and it is therefore highly desirable to use natural stabilizing agents (starch, glucose or plant extracts) and biocompatible solvents such as water.

Although chemists are more likely to think of water as a reaction solvent, it is as a solvent for coatings that it is likely to bring about the most environmental benefits. Aqueous preparations of materials and polymers that can be used in the coating industry are therefore very important. Water based coatings have been around for many years, but new formulations are continually being developed to meet more demanding applications. Replacing an organic solvent with water is not simple and often requires the development of new additives and dispersing agents as well as reformulation of the coating and polymeric material itself. Some of the main advantages and challenges that need to be met in the development of new water based coatings were discussed in Chapter 1. Despite the advantages, discoveries and improvements that have been made in recent years, it is still the consumers who must decide between environmental and technical performance.

3.3 High Temperature, Superheated or Near Critical Water

When water is heated to high temperatures between 100 °C (its usual boiling point) and 374 °C (its critical temperature) in a sealed vessel or under pressure, its properties approach those of supercritical water (SCW) and its hydrogen bond network breaks down.[48–50] In this temperature range, water can be called high temperature, superheated or near critical (NCW). It has a lower polarity (E^N_T, α and dielectric constant), density, viscosity, and surface tension than water at room temperature. However, β (hydrogen bond acceptor ability) remains constant with changing temperature, and diffusivity and specific heat capacity increase. The concentrations of hydronium (H_3O^+) and hydroxide (OH^-) ions also increase as K_w increases with increasing temperature. In general, many organic compounds are more soluble in NCW and inorganic salts are still soluble until the regime close to the critical point is reached. As a rule of thumb, NCW has a E^N_T (polarity) similar to acetone and at higher temperatures becomes completely miscible with toluene. However, comparisons have also been made with methanol and ethanol. Therefore, as a form of water, NCW has been used as an alternative to organic solvents in extractions, recrystallizations, chromatography, and decontamination and waste treatment. Many fields use NCW without being aware of it; temperatures above 100 °C are regularly used in the food and paper and pulp industries. There has also been

Figure 3.18   *p*-Isopropenylphenol synthesis via bisphenol A decomposition in NCW.

extensive interest in the recycling of polymers by depolymerization and regeneration of rubber by devulcanization. Additionally, extensive research has been performed on organic reactions in this unusual solvent.[48,49] For example, *p*-isopropenylphenol can be prepared in the absence of a catalyst through the decomposition of bisphenol A (Figure 3.18).[51] Due to the organic–aqueous nature of the reaction, separation and isolation of *p*-isopropenylphenol could be achieved by cooling the reaction mixture to room temperature, at which point the product precipitates. Maximized yields of the desired product were obtained by performing the reaction at 350 °C for 20 min. Longer reaction times were required at lower temperatures and reaction monitoring was essential to prevent product decomposition, which yielded acetone, further phenol and *p*-isopropyl phenol as by-products.

It should be noted that in such experiments the deionized water is thoroughly degassed using helium to prevent anomalies caused by dissolved gases. Because of the specialized equipment that is generally required for chemistry using NCW and its relationship to SCW, further applications in this area are discussed in Chapter 4.

3.4   Summary and Outlook for the Future

Water is already used on an industrial scale for emulsion polymerizations, hydrodistillations, biochemical transformations and hydroformylation reactions. However, as far as I am aware, it is not being used on a large scale for traditional multi-step organic syntheses. However, pilot scale processes to produce pharmaceutical intermediates on a multi-kilogram scale using tap water as the only solvent have been successful.[52] Therefore, it is only a matter of time before ingenious chemists and chemical engineers in process development laboratories scale up more aqueous phase chemical reactions. This is even more likely given the wide variety of reactions that have been successfully performed in water during the last 20 years; in addition, the price and global availability of water mean that it is probably the ideal green solvent. New reactions will continue to be discovered in water, especially those aided by new methods such as microwave reactors. Much research is likely to occur in the area of bio-sourced chemicals and materials, where enzymes will probably play an important role. In addition to pure aqueous phase chemistries, new technologies that have been developed, including organic–aqueous tunable solvent systems, will allow reduced amounts of VOCs to be used where organic solvents are still necessary because of solubility issues. However, it should also be noted that in many cases reaction rates and yields are significantly improved when reagents are insoluble in the aqueous phase, and in many cases these 'on water' reactions are superior to solvent free approaches. Therefore, if a solvent free approach does not work, and your compounds are not hydrophilic, it is still worth attempting a reaction using water—you may be surprised by the result!

References

1. D. J. Adams, P. J. Dyson and S. J. Taverner, *Chemistry in Alternative Reaction Media*, John Wiley & Sons Ltd., Chichester, 2004.
2. E. Wiebus and B. Cornils, in *Catalyst Separation, Recovery and Recycling*, ed. D. J. Cole-Hamilton and R. P. Tooze, Springer, Netherlands, 2006.
3. B. Cornils and W. A. Herrmann ed., *Aqueous-Phase Organometallic Catalysis*, Wiley-VCH, Weinheim, 2004.
4. C. M. Starks, C. L. Liotta and M. Halpern, *Phase transfer Catalysis: Fundamentals, Applications, and Industrial Perspectives*, Springer, Netherlands, 1994.
5. R. Klein, D. Touraud and W. Kunz, *Green Chem.*, 2008, **10**, 433.
6. N. Asfaw, P. Licence, A. A. Novitskii and M. Poliakoff, *Green Chem.*, 2005, **7**, 352.
7. F. Chemat, M. E. Lucchesi, J. Smadja, L. Favretto, G. Colnaghi and F. Visinoni, *Anal. Chim. Acta*, 2006, **555**, 157.
8. U. M. Lindstrom ed., *Organic Reactions in Water*, Blackwell, Oxford, 2007.
9. D. Dallinger and C. O. Kappe, *Chem. Rev.*, 2007, **107**, 2563.
10. H. C. Hailes, *Org. Process Res. Dev.*, 2007, **11**, 114.
11. C. I. Herrerias, X. Q. Yao, Z. P. Li and C. J. Li, *Chem. Rev.*, 2007, **107**, 2546.
12. C. J. Li, *Chem. Rev.*, 2005, **105**, 3095.
13. U. M. Lindstrom, *Chem. Rev.*, 2002, **102**, 2751.
14. V. T. Perchyonok, I. N. Lykakis and K. L. Tuck, *Green Chem.*, 2008, **10**, 153.
15. D. C. Rideout and R. Breslow, *J. Am. Chem. Soc.*, 1980, **102**, 7816.

16. S. Narayan, H. Muldoon, M. G. Finn, V. V. Fokin, H. C. Kolb and K. B. Sharpless, *Angew. Chem. Int. Ed.*, 2005, **44**, 3275.
17. H. B. Zhang, L. Liu, Y. J. Chen, D. Wang and C. J. Li, *Eur. J. Org. Chem.*, 2006, **869**.
18. N. Shapiro and A. Vigalok, *Angew. Chem. Int. Ed.*, 2008, **47**, 2849.
19. P. G. Cozzi and L. Zoli, *Angew. Chem. Int. Ed.*, 2008, **47**, 4162.
20. S. V. More, M. N. V. Sastry and C. F. Yao, *Green Chem.*, 2006, **8**, 91.
21. R. Ding, K. Katebzadeh, L. Roman, K.-E. Bergquist and U. M. Lindstrom, *J. Org. Chem.*, 2006, **71**, 352.
22. F. J. Waller, A. G. M. Barrett, D. C. Braddock and D. Ramprasad, *Chem. Commun.*, 1997, 613.
23. K. Aplander, R. Ding, U. M. Lindstrom, J. Wennerberg and S. Schultz, *Angew. Chem. Int. Ed.*, 2007, **46**, 4543.
24. W. Sun, H. Wang, C. Xia, J. Li and P. Zhao, *Angew. Chem. Int. Ed.*, 2003, **42**, 1042.
25. N. Komiya, T. Nakae, H. Sato and T. Naota, *Chem. Commun.*, 2006, 4829.
26. Y. M. A. Yamada, T. Arakawa, H. Hocke and Y. Uozumi, *Angew. Chem. Int. Ed.*, 2007, **46**, 704.
27. L. F. Liu, Y. H. Zhang and Y. G. Wang, *J. Org. Chem.*, 2005, **70**, 6122.
28. L. Chen and C. J. Li, *Org. Lett.*, 2004, **6**, 3151.
29. N. E. Leadbeater, *Chem. Commun.*, 2005, 2881.
30. B. A. Roberts and C. R. Strauss, *Acc. Chem. Res.*, 2005, **38**, 653.
31. K. Nakamura and T. Matsuda, in *Organic Reactions in Water*, ed. U. M. Lindstrom, Blackwell, Oxford, 2007.
32. J. M. Broering, E. M. Hill, J. P. Hallett, C. L. Liotta, C. A. Eckert and A. S. Bommarius, *Angew. Chem. Int. Ed.*, 2006, **45**, 4670.
33. E. M. Hill, J. M. Broering, J. P. Hallett, A. S. Bommarius, C. L. Liotta and C. A. Eckert, *Green Chem.*, 2007, **9**, 888.
34. C. Roosen, M. Ansorge-Schumacher, T. Mang, W. Leitner and L. Greiner, *Green Chem.*, 2007, **9**, 455.
35. N. Eghbali and C. J. Li, *Green Chem.*, 2007, **9**, 213.
36. Y. Himeda, *Eur. J. Inorg. Chem.*, 2007, **1**, 3927.
37. Y. Himeda, N. Onozawa-Komatsuzaki, H. Sugihara and K. Kasuga, *Organometallics*, 2007, **26**, 702.
38. S. Mecking, A. Held and F. M. Bauers, *Angew. Chem. Int. Ed.*, 2002, **41**, 545.
39. K. Matyjaszewski, J. Qiu, D. A. Shipp and S. G. Gaynor, *Macromolecular Symposia*, 2000, **155**, 15.
40. S. M. Yu, A. Berkefeld, I. Gottker-Schnetmann, G. Muller and S. Mecking, *Macromolecules*, 2007, **40**, 421.
41. V. Monteil, J. Stumbaum, R. Thomann and S. Mecking, *Macromolecules*, 2006, **39**, 2056.
42. S. Fujii, S. P. Armes, R. Jeans, R. Devonshire, S. Warren, S. L. McArthur, M. J. Burchell, F. Postberg and R. Srama, *Chem. Mater.*, 2006, **18**, 2758.
43. X. Y. Feng, J. L. Zhang, S. Q. Cheng, C. X. Zhang, W. Li and B. X. Han, *Green Chem.*, 2008, **10**, 578.

44. P. Raveendran, J. Fu and S. L. Wallen, *J. Am. Chem. Soc.*, 2003, **125**, 13940.
45. P. Raveendran, J. Fu and S. L. Wallen, *Green Chem.*, 2006, **8**, 34.
46. J. C. Liu, G. W. Qin, P. Raveendran and Y. Kushima, *Chem. Eur. J.*, 2006, **12**, 2132.
47. S. K. Li, Y. H. Shen, A. J. Xie, X. R. Yu, L. G. Qiu, L. Zhang and Q. F. Zhang, *Green Chem.*, 2007, **9**, 852.
48. A. R. Katritzky, D. A. Nichols, M. Siskin, R. Murugan and M. Balasubramanian, *Chem. Rev.*, 2001, **101**, 837.
49. P. E. Savage, *Chem. Rev.*, 1999, **99**, 603.
50. M. Siskin and A. R. Katritzky, *Chem. Rev.*, 2001, **101**, 825.
51. S. E. Hunter, C. A. Felczak and P. E. Savage, *Green Chem.*, 2004, **6**, 222.
52. T. J. Connolly, P. McGarry and S. Sukhtankar, *Green Chem.*, 2005, **7**, 586.

CHAPTER 4
Supercritical Fluids

4.1 Introduction

Supercritical fluids (SCFs) have long fascinated chemists and over the last 30 years this interest has accelerated. There is even a journal dedicated to the subject—the *Journal of Supercritical Fluids*. These fluids have many fascinating and unusual properties that make them useful media for separations and spectroscopic studies as well as for reactions and synthesis. So what is an SCF? Substances enter the SCF phase above their critical pressures (P_c) and temperatures (T_c) (Figure 4.1).[1–3] Some substances have readily accessible critical points, for example T_c for carbon dioxide is 304 K (31 °C) and P_c is 72.8 atm, whereas other substances need more extreme conditions. For example T_c for water is 647 K (374 °C) and P_c is 218 atm. The most useful SCFs to green chemists are water and carbon dioxide, which are renewable and non-flammable. However, critical data for some other substances are provided for comparison in Table 4.1. In addition to reactions in the supercritical phase, water has interesting properties in the near critical region and carbon dioxide can also be a useful solvent in the liquid phase. Collectively, carbon dioxide under pressurized conditions (liquid or supercritical) is sometimes referred to as dense phase carbon dioxide.

The *critical point* of an SCF represents the highest temperature and pressure at which the substance can exist as a vapour and liquid in equilibrium. At the *triple point*, the solid, liquid and gas phases coexist. The gas–liquid coexistence curve is known as the *boiling curve*. If we move upwards along the boiling curve, increasing temperature and pressure, then the liquid becomes less dense due to thermal expansion and the gas becomes denser as the pressure rises. Eventually, the densities of the two phases converge and become identical, the distinction between gas and liquid disappears, and the boiling curve comes to an end at the critical point. This transition can be observed using a high-pressure view cell.

RSC Green Chemistry Book Series
Alternative Solvents for Green Chemistry
By Francesca M. Kerton
© Francesca M. Kerton 2009
Published by the Royal Society of Chemistry, www.rsc.org

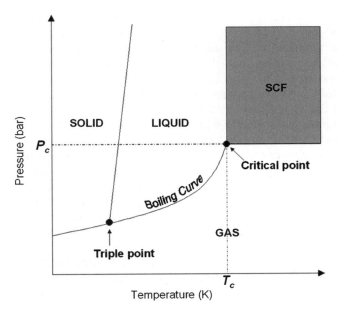

Figure 4.1   Single-component phase diagram highlighting the supercritical fluid (SCF) region and the critical point.

Table 4.1   Critical point (T_c and P_c) and critical density for selected compounds.

Substance	T_c/K	P_c/atm	$\rho/g\,ml^{-1}$
CHF_3	299.3	46.9	0.528
CH_4	190.5	41.4	0.162
C_2H_4	282.3	50.5	0.215
C_2H_6	305.2	48.2	0.203
CO_2	304.1	72.8	0.469
H_2O	647.1	218.3	0.348
CH_3CH_2OH	513.9	60.6	0.276
Xe	289.7	58	1.110

Video footage of the disappearing meniscus can be seen at several sites on the Internet.[4]

Both batch and continuous-flow reactors have been used for reactions in SCF. Batch reactors can be readily equipped with a suitable window to assess homogeneity of the reaction mixture and are widely used in academic research. These windows can also be used for spectroscopic analysis such as FT-IR. Other methods for assessing the homogeneity of mixtures have recently been reported; these include a piezoelectric sensor.[5] Such techniques reduce potential human errors that are possible using methods that involve observations through high-pressure windows. It is worth noting that the homogeneity of a reaction mixture can change during the course of a reaction. It should also be emphasized that as reagents or co-solvents are added to an SCF, its properties can change significantly, and the phase diagram for the reaction mixture may differ considerably from that of the pure substance.

One of the main differences between SCFs and conventional solvents is their compressibility. No distinct gas or liquid phase can exist above the critical point, and the SCF phase has a unique combination of properties from both gas and liquid states (Table 4.2). At liquid-like densities SCFs exhibit low viscosity and high diffusion rates, like a gas. Conventional solvents require very large pressure changes to vary their density, whereas the density of an SCF changes significantly upon increasing pressure. Solubility in an SCF is related to density, therefore this medium has the added benefit of being tuneable, and hence the solubility of species can be directly controlled. Purification or reaction quenching can thus be achieved by reducing solvent density and precipitating the product. Varying the density can also affect the selectivity and outcome of some chemical reactions.

The extensive interest in SCFs is partially due to the additional benefits that SCFs offer besides being environmentally benign (Table 4.3).[7,8] Many of these result from SCFs having physical properties intermediate between those of gases and liquids.

It should be noted that on an industrial scale, reactions or other processes in SCF media are not new. Many industrial reactions developed in the early part of the twentieth century are actually conducted under supercritical conditions of either their product or reagent including ammonia synthesis (BASF, 1913), methanol synthesis (BASF, 1923) and ethylene polymerization (ICI, 1937).

Table 4.2 Comparison of typical diffusivities, viscosities and densities of gaseous, supercritical and liquid phases.[6]

Property	Gas	SCF	Liquid
Diffusivity/$cm^2 s^{-1}$	10^{-1}	2×10^{-4}	5×10^{-6}
Viscosity/$g\,cm^{-1} s^{-1}$	10^{-4}	2×10^{-4}	10^{-2}
Density/$g\,cm^{-3}$	10^{-3}	$0.1–0.9(CO_2\ 0.4)$	1.0

Table 4.3 Summary of advantageous properties of SCFs in general.

High solubility of any reacting gases means that hydrogenations and other reactions
   involving gaseous reagents are enhanced in their selectivity and energy requirements
Rapid diffusion
Weakening of solvation around the reacting species
Reduction of cage effects in radical reactions
Solvent is easily removed owing to its 'zero' surface tension, leaving the product in an
   easily processable, clean and solvent-free form
Recyclability, and therefore near zero waste production

4.2 Chemical Examples

4.2.1 Supercritical and Liquid Carbon Dioxide

In many cases, carbon dioxide is seen as the most viable supercritical solvent. It is inexpensive and can be obtained as a by-product of fermentation and combustion. It is non-toxic and not a VOC. It is non-flammable and relatively inert, especially when compared with other alternatives. It can react with nucleophiles (*e.g.* carbamic acid formation from amines), although this can be reversible, and subsequently exploited synthetically.[9,10] Carbon dioxide also provides many chemical advantages that enhance its green credentials by reducing waste.[11] For example, it cannot be oxidized and therefore oxidation reactions using air or oxygen as the oxidant have been intensively investigated. Also, it is inert to free radical chemistry, in contrast to many conventional solvents. This has led to much research into polymerizations initiated by free radicals.[8] There are also a number of practical advantages associated with the use of supercritical carbon dioxide ($scCO_2$) as a solvent. Product isolation to total dryness is achieved by simple evaporation and could prove useful in the final steps of pharmaceutical syntheses where even trace amounts of solvent residues are considered problematic. Given the critical point of carbon dioxide, most processes reported to date have been conducted in a pressure regime of 100–200 bar. The potential danger of such conditions should never be ignored and safety precautions should be taken for all experiments. The advantages and disadvantages of $scCO_2$ as a solvent are listed in Table 4.4.

4.2.1.1 Solubility in Supercritical Carbon Dioxide

All gases are miscible with SCFs. This is particularly important for catalysis in SCFs and particularly $scCO_2$. For example, the concentration of hydrogen in a supercritical mixture of hydrogen (85 bar) and carbon dioxide (120 bar) at 50 °C is 3.2 M, whereas the concentration of hydrogen in THF under the same pressure is merely 0.4 M.[7] Therefore, there is potential for much improved chemical processes where gaseous reagents have traditionally been used in the solution phase. An additional feature of SCFs, which enhances solubility at moderate densities (near the critical point) is solute-solvent clustering (Figure 4.2).[12]

In addition to gases, other reagents including low molecular weight organic compounds, *e.g.* cyclohexene and caffeine, possess good miscibility or solubility in SCFs. It is important to assess the solubility and phase behaviour of reactants as the reaction might be occurring as a 'solvent free' process under an atmosphere of carbon dioxide and not actually accessing the full benefits of using $scCO_2$. The traditional method for obtaining solubility data for substances in SCFs is cloud point data. Temperature and pressure are varied for a solvent–solute system and a graph is acquired that indicates when the substance falls out of solution and forms 'clouds'.

There are a number of methods that can be applied to increase the solubility of insoluble materials (Table 4.5). Cheaper, more sustainable approaches to

Table 4.4   Summary of properties of scCO$_2$ as a solvent.

	Advantages	*Disadvantages*
Environmental and safety	No liquid waste/solvent effluent Non-flammable Non-toxic to the environment/ personnel Available cheaply and in >99.9% pure form	Involves high pressures
Reaction and process	Low viscosity	Equipment costs; pressure vessels are required
	Gas miscibility	Heat transfer limitations; faster reaction rates can be problematic for particularly exothermic reactions
	Simple product isolation by evaporation to 100% dryness	Weak solvent; relatively non-polar, co-solvents or modification of reagents needed to improve solubility, but many low MWt non-polar compounds are soluble
	Range of processing techniques available, such as RESS	Reacts in the presence of good nucleophiles
	High diffusion rates offer potential for increased reaction rates	Misplaced technophobia
	Density can be varied to control reagent/product solubility, 'tunable' solvent	
	Relatively inert and non-oxidizable	

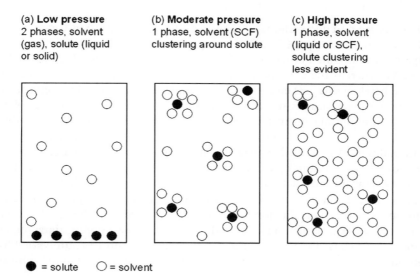

(a) **Low pressure**
2 phases, solvent
(gas), solute (liquid
or solid)

(b) **Moderate pressure**
1 phase, solvent (SCF)
clustering around solute

(c) **High pressure**
1 phase, solvent
(liquid or SCF),
solute clustering
less evident

● = solute   ○ = solvent

Figure 4.2   Schematic representation of solute–solvent clustering in an SCF, compared with liquid-phase solvation and lack of solvation in the gas phase.

Table 4.5 Methods that can be used to overcome the limited solvating power of $scCO_2$.

Increasing the bulk density of the SCF	Simple but not always desired, as higher pressures mean higher costs.
Addition of a co-solvent	Modifiers (*e.g.* MeOH) can be added to increase or decrease polarity. However, the more modifier that is added, the further $scCO_2$ moves away from being the ideal green solvent. Reagents themselves may also in effect act as co-solvents.
Biphasic syntheses	Carbon dioxide is finding increasing use in combination with other green solvents, including ionic liquids and water.
Modification of the solute	Organic fluorocarbons, siloxanes and related compounds show greater solubility in $scCO_2$ compared with the corresponding hydrocarbons. These compounds are collectively known as 'CO_2-philes'. The effect of the increased solubility of fluorocarbon species has been used in the design of surfactants, chelating agents, and ligands in order to enhance the solubility of polymers, metals and catalysts respectively.[16]

solubility enhancement are likely to have a significant impact in the future. Polyether–carbonate copolymers made from propylene oxide and carbon dioxide using an aluminium catalyst were designed as CO_2-philes after analysing the thermodynamic factors affecting solubility in carbon dioxide.[13] The C–O–C backbone of the polymer is highly flexible and has only weak solute–solute interactions. Additionally, there are favourable interactions of the carbonyl group with carbon dioxide. Beckman was awarded a US EPA award in 2002 for this work. Other new CO_2-philes include peracetylated sugars,[14] and most recently, very stable carbon dioxide-in-water emulsions have been created using the relatively cheap and innocuous protein β-lactoglobulin as the emulsifier.[15] These latter two examples of carbon dioxide soluble species and emulsifying agents based on natural materials show great promise for the future of carbon dioxide in the processing of a much wider of range of materials than previously thought.

Two classes of polymeric materials, amorphous fluoropolymers and silicones, are the only commercially available polymers to exhibit appreciable solubility in $scCO_2$ at readily accessible temperatures and pressures (Figure 4.3). It has been proposed that this results from a special interaction between fluorine and carbon dioxide due to the polarity of both species. Silicones are also thought to dissolve because they have weak intermolecular interactions and flexible backbones.

Because of their widely recognized solubility in $scCO_2$, fluoropolymers have become extensively used as modifiers in this medium (Figure 4.4). They have formed the basis of surfactants for dispersion polymerizations and water microemulsion formation, as extractants for metals and as modifiers to dissolve insoluble organic reagents, *e.g.* radical initiators and tin reagents.

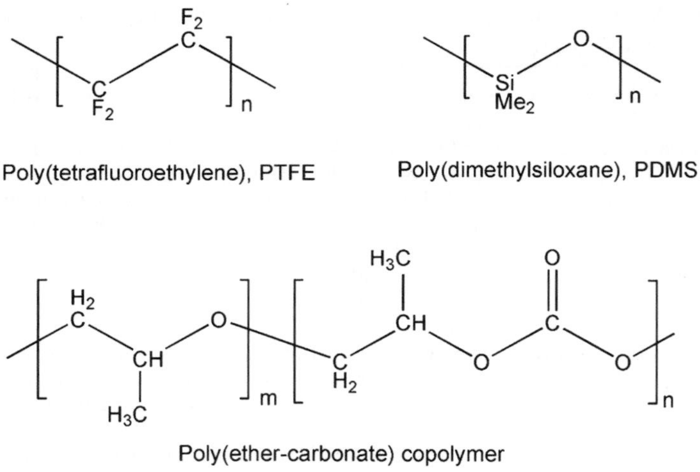

Poly(tetrafluoroethylene), PTFE   Poly(dimethylsiloxane), PDMS

Poly(ether-carbonate) copolymer

Figure 4.3   Polymers that are soluble in scCO$_2$.

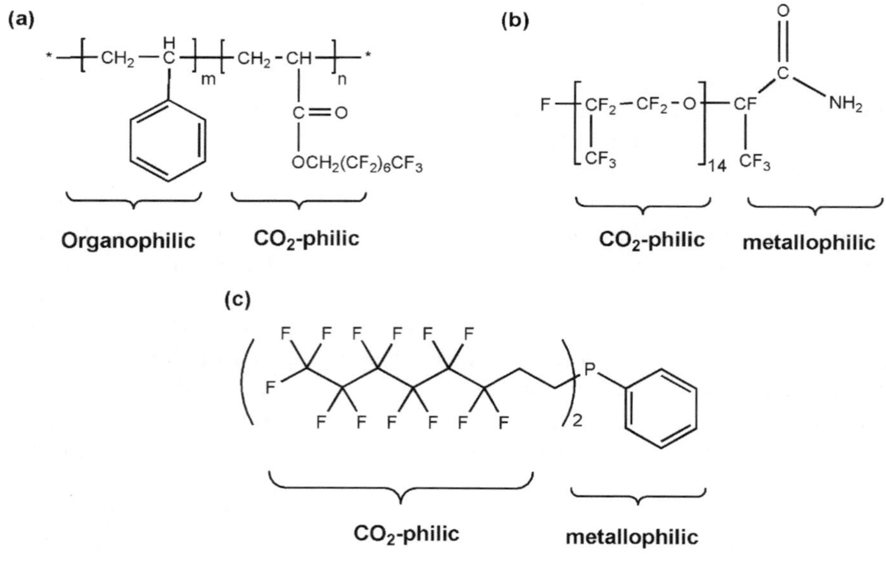

Figure 4.4   Some fluoropolymer derived materials used in scCO$_2$ technologies: (a)
copolymer used as stabilizer in emulsion polymerizations of styrene; (b)
end functionalized polymer used in metal extraction studies; (c) a ligand
used for homogeneous catalysis in scCO$_2$.

4.2.1.2   Extraction

Supercritical carbon dioxide is widely used in supercritical fluid extraction
(SFE) and supercritical fluid chromatography (SFC).[1] It is a good extraction

solvent as it is chemically pure, non-toxic, non-flammable, non-polar, stable, colourless, odourless and tasteless. Importantly, it is easily removed and highly selective. Upon extraction, further processing is possible; for example, the scent can be impregnated into a material for slow release. Industrially, carbon dioxide has been used in the beverage, food and flavour, and cosmetics industries. This is partially because significant value is added as products that are processed using carbon dioxide can be labelled natural and environmentally friendly. Some of the advantages and disadvantages of SFE compared with other extraction techniques are outlined in Table 4.6.

The use of $scCO_2$ in extraction and chromatography has recently been reviewed.[17–19] Equipment is commercially available for both processes, on a large and small scale, and the two techniques can be hyphenated. Essentially, for static extractions, a pressure vessel and carbon dioxide pump are required, but most extractions are performed under flow conditions that require an additional back-pressure regulator and flow meter. It should be noted that

Table 4.6 Comparison of SFE with some other commonly used extraction techniques.

	Advantages	*Disadvantages*
SFE	Low-temperature extraction results in minimal degradation of volatile compounds.	Very high capital installation costs.
	Higher product yields than with steam distillation.	High running costs.
	Spent material undamaged, unlike steam distillation/solvent extraction.	Requires technically skilled operators.
		Not suitable for wet raw materials.
		Lower product yield than solvent extraction.
Steam distillation	Low capital running costs	Unpredictable degradation of some groups of compounds.
	Applicable to most essential oils, fragrances and flavour compounds.	Cleaning between products can be difficult.
	Designs available to suit all volumes.	Extraction of further products from residue can be difficult due to high moisture level.
Solvent extraction	Non-selective; wide spectrum of compounds extracted simultaneously (that can be a disadvantage too).	Most solvent residues must be monitored and tightly controlled.
	Extraction carried out at various temperatures and pressures.	Most commonly used solvents are highly flammable and possibly toxic.
	Solvents can be readily removed at atmospheric or reduced pressure.	Waste has little or no value.

liquid carbon dioxide can also be used for extractions and in some cases, *e.g.* extraction of limonene from orange peel as a teaching laboratory experiment, no special equipment is required.[20] In contrast, the equipment for SFC is much more expensive and complicated but this technique is currently finding renewed popularity as a separation and analytical tool because of the speed at which separations can be performed.

A complete review of all extractions performed using $scCO_2$ or liquid carbon dioxide is beyond the scope of this book. However, it should be noted that in most cases, the technique is complementary to water based extraction methods. For example, SFE can successfully be used to extract valuable waxes and higher molecular weight sesquiterpenes that are not water soluble.[21–23] This offers the opportunity to perform fractional, green extractions. Also, given the number of variables in optimizing a SFE process, experimental design can play an important role.[22] This was used in the extraction of valuable wax products from wheat straw (an agricultural by-product) and was scaled up to a > 75 kg level at an industrial extraction plant. Interestingly, the quality of the wax was strongly dependent on the extraction conditions and the use of a co-solvent (ethanol) was deleterious and led to a complete loss of selectivity.

4.2.1.3   *Chemical Synthesis*

A large and continually expanding list of reactions has been performed in $scCO_2$.[16,24–27] Many of these reactions, including hydrogenations, hydro-formylations and oxidations, make use of the unique properties of SCFs such as gas miscibility. Other reactions show increased selectivity due to special interactions of the solvent with the substrate (Diels–Alder reactions and sulfur oxidation), or an increased potential towards industrial development due to process intensification (continuous flow reactors) and reduced post reaction purification *e.g.* catalyst separation. A simple laboratory reaction set-up for $scCO_2$ work is shown in Figure 4.5. Many reactors are equipped with high-pressure windows to view the ongoing reactions.

Some organic reactions that have been performed in $scCO_2$ are shown in Figure 4.6. Two reactions that have shown very unusual pressure-dependent selectivities are Lewis acid catalysed Diels–Alder reactions and diastereoselective sulfur oxidation.[28–30] In general, the most dramatic changes in reactivity and selectivity are seen around the critical density of the solvent. Further reactions will likely show similar pressure and density dependent selectivity in the future, although, the best selectivities are not always seen in carbon dioxide. In the Henry reaction, neat (solvent free) conditions afforded the highest conversions and selectivity. In the Diel–Alder reaction, it is thought that the triflate anion, $CF_3SO_3^-$, assists in solubilizing the Lewis acid. A similar effect was observed in the iron-mediated oxidative polymerization of pyrrole in $scCO_2$ using iron tri-flate, $Fe(CF_3SO_3)_3$.[31] A review of homogeneously catalysed reactions in super-critical and liquid carbon dioxide has recently been published.[32] Reactions studied include aldol reactions, carbonylations, cyclizations, epoxidations,

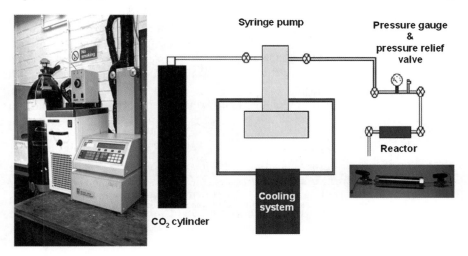

Figure 4.5 Schematic diagram of typical scCO$_2$ laboratory set-up for reactions. Inset left, cylinder, circulating chilling unit and syringe pump. Inset right, typical pressure vessel.

esterifications, carbon–carbon cross coupling reactions, hydrogenations, hydroformylations and polymerizations. By far the most extensively studied of these are hydrogenations and hydroformylations because of the high solubility of reagent gases in scCO$_2$ compared to conventional organic solvents. Second to these are palladium catalysed carbon–carbon coupling reactions because of their versatility in building up complex molecules.

Some of the seminal work in the area of catalysis in scCO$_2$ was performed by Noyori and co-workers.[7,33] They discovered that relatively simple ruthenium complexes could be used to catalyse the conversion of carbon dioxide to formic acid. This reaction took advantage of the miscibility of hydrogen gas in scCO$_2$, which also acted as a reagent and not just a solvent for the process. This work was then extended by researchers at the Los Alamos National Laboratory to industrially important asymmetric hydrogenation reactions.[34] In this case, the solubility of the asymmetric rhodium catalyst was enhanced by using the tetra-kis(3,5-bis(trifluoromethyl)phenyl)borate anion. The enantioselectivities achieved in scCO$_2$ were competitive with control reactions performed in conventional solvents. Many more hydrogenation studies have been performed, including those using biphasic approaches and heterogeneous catalysts that will be discussed later. In addition to hydrogenation reactions, rhodium and manganese metal complexes have been used to catalyse a range of homogeneous hydroformylation reactions in carbon dioxide.[35–39] However, for hydroformylation, other alternative solvent systems (*e.g.* water, carbon dioxide expanded liquids) have so far given superior results, especially when the additional energy costs of pressurization are taken into account.

If we consider the case of palladium-mediated carbon–carbon bond formation in scCO$_2$, initially the main problem was the insolubility of reagents that

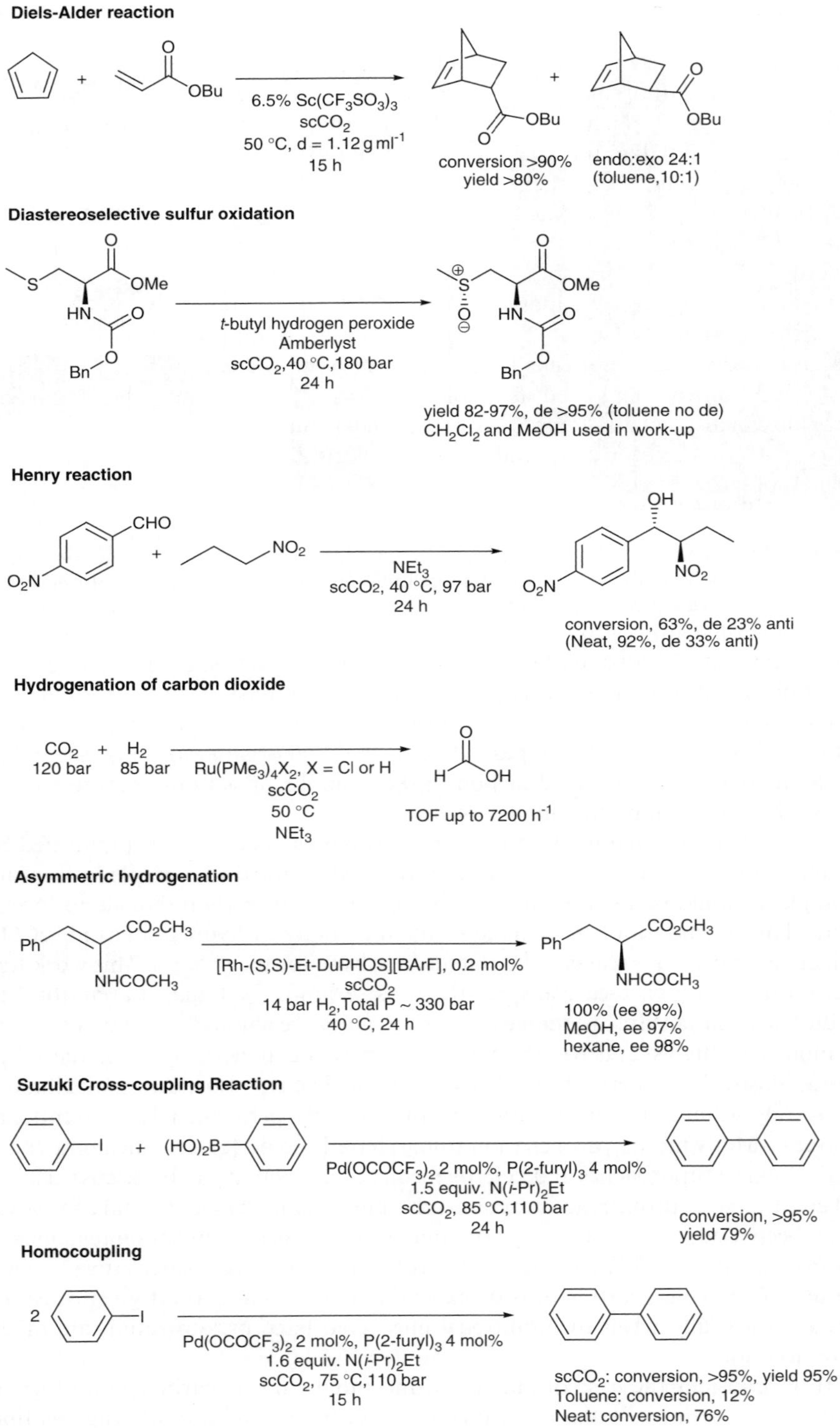

Diels-Alder reaction

6.5% Sc(CF₃SO₃)₃
scCO₂
50 °C, d = 1.12 g ml⁻¹
15 h

conversion >90%
yield >80%

endo:exo 24:1
(toluene,10:1)

Diastereoselective sulfur oxidation

t-butyl hydrogen peroxide
Amberlyst
scCO₂,40 °C,180 bar
24 h

yield 82-97%, de >95% (toluene no de)
CH₂Cl₂ and MeOH used in work-up

Henry reaction

NEt₃
scCO₂, 40 °C, 97 bar
24 h

conversion, 63%, de 23% anti
(Neat, 92%, de 33% anti)

Hydrogenation of carbon dioxide

CO₂ + H₂
120 bar 85 bar

Ru(PMe₃)₄X₂, X = Cl or H
scCO₂
50 °C
NEt₃

TOF up to 7200 h⁻¹

Asymmetric hydrogenation

[Rh-(S,S)-Et-DuPHOS][BArF], 0.2 mol%
scCO₂
14 bar H₂,Total P ~ 330 bar
40 °C, 24 h

100% (ee 99%)
MeOH, ee 97%
hexane, ee 98%

Suzuki Cross-coupling Reaction

Pd(OCOCF₃)₂ 2 mol%, P(2-furyl)₃ 4 mol%
1.5 equiv. N(i-Pr)₂Et
scCO₂, 85 °C,110 bar
24 h

conversion, >95%
yield 79%

Homocoupling

Pd(OCOCF₃)₂ 2 mol%, P(2-furyl)₃ 4 mol%
1.6 equiv. N(i-Pr)₂Et
scCO₂, 75 °C,110 bar
15 h

scCO₂: conversion, >95%, yield 95%
Toluene: conversion, 12%
Neat: conversion, 76%

Figure 4.6 Some organic reactions studied in scCO₂.

were typically used including palladium complexes, such as $Pd(OAc)_2$ and $Pd(PPh_3)_4$, and inorganic bases (*e.g.* K_3PO_4). This was overcome by using fluorinated ligands such as $(C_6F_{13}CH_2CH_2)_2PPh$ and $P[3,5-(CF_3)_2C_6H_3]_3$,[40,41] or a fluorinated palladium source.[42] The insolubility of the bases was overcome by using an organic base such as diisopropylethylamine. However, in general, superior results for these reactions, particularly Suzuki cross couplings, can be obtained using water as the solvent, especially if used in conjunction with microwave heating.[43] In some cases, however, evidence has shown that couplings are more effective in $scCO_2$ than in toluene or under solvent free conditions.[44] The use of $scCO_2$ has subsequently been applied to many types of palladium catalysed reactions including C–N bond formation.[45]

In addition to being used as a solvent or as a reagent and solvent, carbon dioxide can also act as a temporary protecting group.[9,10] Carbon dioxide inserts into N–H bonds of RNH_2 and R_2NH molecules, sometimes reversibly and sometimes irreversibly. If the process is reversible, it can be exploited. This strategy has been used successfully in a ruthenium catalysed ring-closing metathesis reaction of α,ω-alkenes containing a secondary amine in the backbone,[9] and rhodium catalysed hydroaminomethylation reactions.[10] In the latter reaction, a cyclic amide product was produced in conventional solvents rather than the cyclic amine that was formed in $scCO_2$. This atom-efficient approach to the protection of amine groups might well be applicable to other organic reactions.

Heterogeneous catalytic reactions in $scCO_2$. The use of heterogeneous catalysts in combination with $scCO_2$ is an alternative solvent success story, having led to commercialization of a hydrogenation process that will be discussed in Chapter 10.[46] This type of continuous hydrogenation process has recently been used for the conversion of a pharmaceutical intermediate, and more advances in this area are therefore expected soon.[47] In addition to hydrogenations, reactions that have been studied in this manner include alkylations, aminations, etherifications, esterification and oxidations.[48] The use of heterogeneous catalysts with $scCO_2$ was recently the subject of a short review.[49] On a laboratory scale the use of flow reactors permits smaller reaction vessels and continuity, leading to a safer process. Additionally, online real-time IR monitoring is possible. These reactors are simple to construct and modify, and possess excellent mass- and heat-transfer properties. By manipulating the phase behaviour in many of the processes, particularly where water is formed as a by-product, the organic products can be separated easily and cleanly. An interesting example is the conversion of water soluble levulinic acid (a biorefinery platform chemical) into γ-valerolactone (Figure 4.7), which can be separated from an aqueous phase by exposure to carbon dioxide. In this example, any unreacted levulinic acid could be easily recycled and pure γ-valerolactone could be isolated even when the hydrogenation reaction was incomplete.[50] Continuous-flow reactors can also yield tunable reactions. A solid acid catalyst has been used to catalyse the reaction

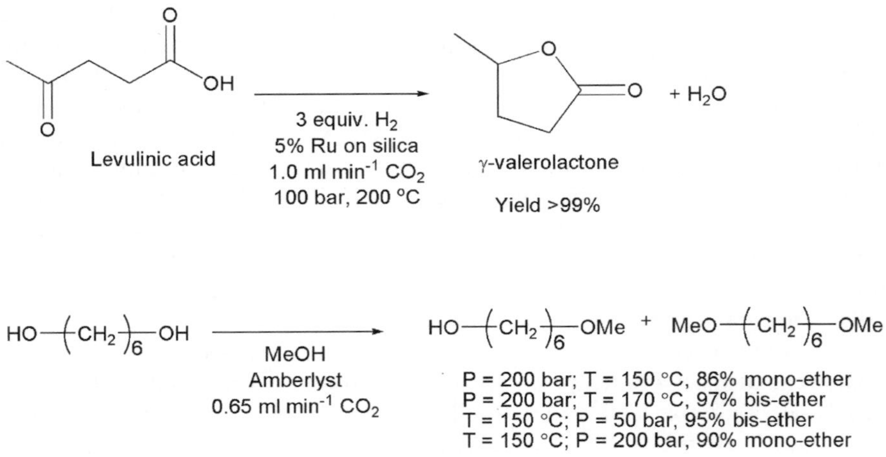

Levulinic acid → γ-valerolactone + H₂O

3 equiv. H$_2$
5% Ru on silica
1.0 ml min^{-1} CO$_2$
100 bar, 200 °C

Yield >99%

HO—(CH$_2$)$_6$—OH → HO—(CH$_2$)$_6$—OMe + MeO—(CH$_2$)$_6$—OMe

MeOH
Amberlyst
0.65 ml min^{-1} CO$_2$

P = 200 bar; T = 150 °C, 86% mono-ether
P = 200 bar; T = 170 °C, 97% bis-ether
T = 150 °C; P = 50 bar, 95% bis-ether
T = 150 °C; P = 200 bar, 90% mono-ether

Figure 4.7   Recent reactions catalysed by heterogeneous catalysts in scCO$_2$.

between 1,6-hexanediol and simple alcohols such as methanol.[51] The selectivity of the reaction was found to be dependent on the density of the SCF phase and could be controlled by adjusting either temperature or pressure (Figure 4.7).

Biphasic carbon dioxide–aqueous and carbon dioxide–ionic liquid systems. ScCO$_2$ and another green solvent, either water or an RTIL, have been used together to perform catalytic reactions. For example, Beckman and co-workers have looked at the direct reaction of hydrogen and oxygen to give hydrogen peroxide (a widely used green oxidant) under biphasic carbon dioxide–water conditions.[52–54] In some cases, they employed a CO$_2$-philic palladium catalyst. The aqueous hydrogen peroxide generated can then be used for green, biphasic alkene epoxidation reactions.

If we consider hydrogenation reactions performed under aqueous–scCO$_2$ biphasic conditions, two options have been explored. The use of a water soluble catalyst, *e.g.* RuCl$_3$–P(C$_6$H$_4$SO$_3$Na)$_3$, gives a potentially recyclable aqueous catalytic phase.[55] The other option is to use a fluorophilic catalyst. This leads to an inverted scCO$_2$–aqueous biphasic system.[56,57] In the study of enantioselective hydrogenations of polar substrates (Figure 4.8), total turnovers of between 1000 and 2000 were achieved and little contamination of the organic phase with rhodium was observed. In such a system, the supercritical phase is never depressurized and therefore this could lead to significant energy savings.

The use of ionic liquids with scCO$_2$ has recently been reviewed.[58] More information on ionic liquids can be found in Chapter 6. However, their use in biphasic catalysis with scCO$_2$ is discussed here. They have been used most extensively for hydrogenation and hydroformylation reactions.

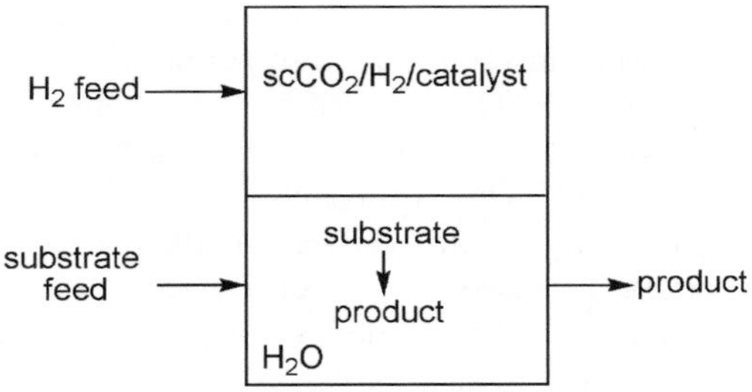

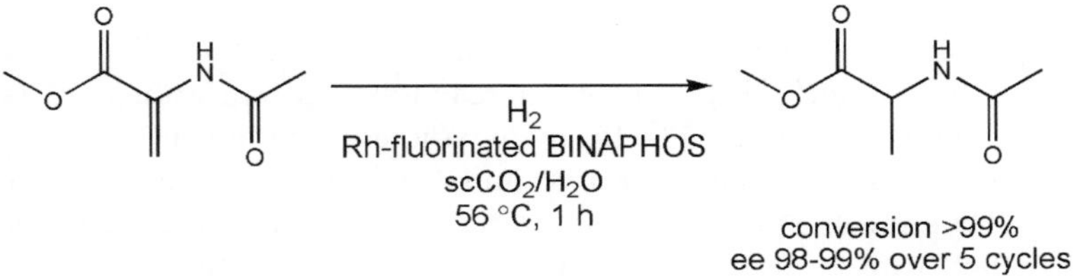

Figure 4.8 Inverted scCO$_2$–aqueous biphasic enantioselective hydrogenation reaction.

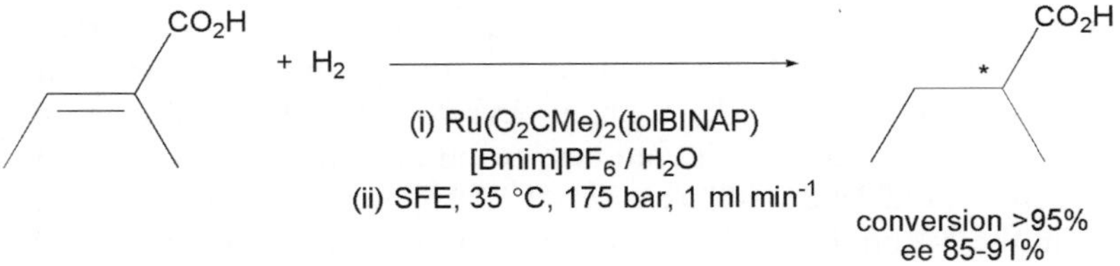

Figure 4.9 Hydrogenation in a biphasic ionic liquid–scCO$_2$ system.

In 2001, Baker and Tumas reported the use of 1-butyl-3-methylimidazolium hexafluorophosphate ([Bmim][PF$_6$]) and rhodium or ruthenium complexes as phase-separable and recyclable hydrogenation catalysts for alkenes and carbon dioxide (in the presence of dialkylamines).[59] Excellent conversions and recyclability were possible upon optimizing the reaction conditions and choice of metal complex. More recently, carbon dioxide has been hydrogenated using a task-specific ionic liquid and a heterogeneous ruthenium catalyst.[60] Although this multiphasic system was not reported as a supercritical reaction, the total pressure that provided the highest turnover frequency was 180 bar at 60 °C. The increased rate at higher pressures was attributed to a concentration effect. Asymmetric hydrogenations have also been reported using chiral ruthenium catalysts in [Bmim][PF$_6$] (in the presence of water or an alcohol as a co-solvent).[61] The products can be extracted using scCO$_2$ and the catalyst phase recycled four times with no significant drops in conversion or enantioselectivity (Figure 4.9).

In rhodium catalysed hydroformylation reactions, conversions achieved using a biphasic system were lower than those achieved in pure ionic liquid: 40% in [Bmim][PF$_6$]–scCO$_2$ *vs* 99% in [Bmim][PF$_6$] alone.[62] However, the selectivity of linear to branched isomer was reversed and therefore these results were highly significant. This approach led to the development of a continuous-flow system for hydroformylation of alkenes, and under careful control the system could be used for several weeks without any visible sign of catalyst degradation.[63] It should be noted that biocatalysts have also been used and recycled using biphasic ionic-liquid–carbon dioxide approaches.[64]

Biocatalysis in scCO$_2$. A wide range of biocatalytic reactions have been performed in scCO$_2$ including hydrolysis reactions, esterifications, carboxylations and polymerizations.[65,66] In these studies, one must be aware that carbon dioxide is potentially reactive and can form carbamates within the enzyme structure, or can react with water to form carbonic acid. The first reaction may lead to decreases in selectivity due to changes in the tertiary structure of the enzyme. The second may affect the pH of the reaction mixture, which in turn could affect the stability and activity of the enzyme. Pressure and temperature can also significantly affect the activity and selectivity of enzymes in scCO$_2$. Biocatalysis in scCO$_2$ could be particularly important in the transformation of bio-feedstocks. For example, the supported lipase enzyme (Novazyme 435) can be used for the quantitative esterification of lavandulol using the naturally sourced acyl donor, acetic acid (Figure 4.10). In this and many biocatalytic studies, reaction temperatures must be kept below a threshold level, in this case 60 °C, to prevent catalyst degradation. Also, in these and other kinetic resolution reactions, enantioselectivity is reduced by increasing the reaction temperature.

4.2.1.4 Materials Synthesis and Modification

The application of scCO$_2$ to the synthesis and modification of well-defined polymers has enormous potential and as such has been extensively investigated.[24] One of the earliest reactions studied was fluoropolymer synthesis.[67]

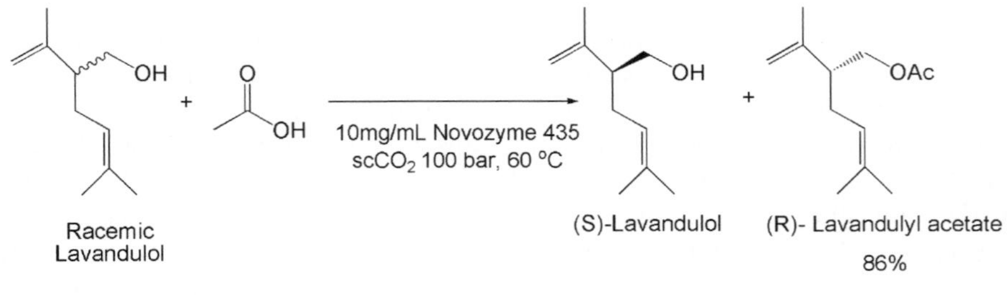

Figure 4.10   Biocatalytic esterification of bio-sourced chemicals.

Fluoropolymers could not be prepared in hydrocarbon solvents and earlier routes to them had employed ozone-depleting chlorofluorocarbon (CFC) solvents. The CO_2-philic nature of both the monomers and the resulting polymers allowed a homogeneous polymerization reaction to be performed. DuPont now uses related technology in the manufacture of Teflon[TM].[68] However, many other polymers, including polymethylmethacrylate (PMMA) and polystyrene (PS), are insoluble in scCO_2, although their monomers are carbon dioxide soluble. Therefore, in order to perform a wider range of polymerizations in this alternative solvent, CO_2-philic/phobic stabilizers or surfactants (Figure 4.4) were developed that could be used to stabilize the growing PMMA and PS chains in scCO_2.[69] These stabilizers are often block copolymers of perfluorooctylacrylate and the monomer to be polymerized. Since these initial studies, the area has grown tremendously and has been extensively reviewed (Figure 4.11).[8,68,70–72] The development of CO_2-philic surfactants has also impacted other areas: for example, they are used in new dry cleaning technologies that avoid the use of perchloroethylene (perc). Nonetheless, because of improvements in the efficiency of traditional perc based dry cleaning units, which reduced solvent usage by two-thirds, there has been limited uptake of this new, expensive technology.

In addition to the preparation of homopolymers and copolymers in scCO_2, extensive processing techniques are available to materials chemists. For example, polymer impregnation is possible due to the high diffusivity of SCFs, which allows them to penetrate throughout the matrix, forming a homogeneous layer of the active compounds of interest. Such technology has been used in the impregnation of drugs in patches or medical devices, preservatives and aroma in food products, and dyeing of textiles, including polyesters. This technology has the potential to substitute classical aqueous dyeing and thereby avoid the related water pollution problems. Materials can also be impregnated with or reacted in the presence of CO_2-philic metal complexes that can be subsequently reduced or thermally decomposed to give metal nanoparticles. For example, an organometallic silver complex has been used recently to give a silver–PMMA composite material.[73] However, it should be noted that an important new method has shown that the metal precursors do not need to be soluble in scCO_2

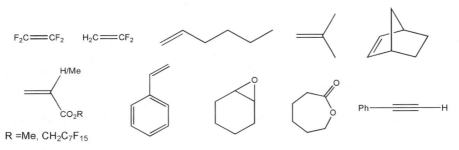

Figure 4.11   Selection of monomers polymerized in scCO_2.

in order to form metallic nanoparticles,[74] and that the plasticizing and swelling behaviour that carbon dioxide can induce in polymers is sometimes sufficient to enable impregnation and subsequent particle growth. The use of SCFs in the preparation of nanoparticles and nanomaterials in general has recently been reviewed.[75] The solubility of metal complexes in scCO$_2$ can also be applied to etching of metal surfaces, which is an important process for device fabrication in the microelectronics industry. For example, it has been demonstrated that fluorinated β-diketones can be used to remove and etch copper from surfaces and it has been proposed that this could be useful in the development of a 'dry' carbon dioxide based process for polishing or smoothing surfaces.[76]

Supercritical drying and particle formation processes are also important areas in scCO$_2$ based materials chemistry. Some of the particle processing methods available, such as precipitation with compressed antisolvent (PCA), are shown schematically in Chapter 9.

Using SCF processes such as these, polymers and inorganic materials have been formed into films, fibres and spherical particles. For example, mesoporous silicate films and mesoporous silica hollow spheres have both been recently prepared using scCO$_2$ based technologies.[77,78]

4.2.2   Supercritical Water and Near Critical Water

A wide and increasing range of synthetic reactions have been performed in near critical water (NCW; around 275 °C, 60 bar) and supercritical water (SCW; around 400 °C, 200 bar).[79,80] As described in Chapter 3, the solvent properties of NCW are similar to those of a polar organic solvent such as acetone. As K_w (the ion product of water) increases with temperature, [H$_3$O$^+$] and [OH$^-$] concentrations are high compared to room temperature, and this leads to many of the interesting properties of NCW and SCW. However, NCW is less corrosive than SCW and requires lower temperatures and pressures. Recently research in this area has increased, especially in extractions and microwave assisted syntheses. Then again, there are some advantages that SCW has over NCW. For example, as with all supercritical fluids, above the critical point of water, gases are highly miscible and this has been used for oxidation chemistry in SCW.

4.2.2.1   Extraction and Analytical Chemistry

Because of its corrosive nature and tendency to degrade (oxidize) organic compounds, SCW is rarely used in the extraction of natural products. However, NCW, and even room temperature pressurized water, have excellent properties for this purpose. In combination, they may be able to selectively extract a range of compounds with interesting biological activity just by gradually increasing the temperature of the extraction medium. They are complementary techniques to SFE as non-polar fractions are not generally extracted with these methods, and they often selectively extract highly valued oxygenated species. Data from some studies in this area are presented in

Table 4.7 Some examples of natural product extraction using NCW.

Plant	Optimized NCW conditions	Comparative methods	NCW benefits
Laurel leaves[81]	150 °C, 50 bar, 2.0 ml min^{-1}	VOC extraction (CH_2Cl_2) and hydrodistillation	Shorter extraction times; better quality oil; more selective; lower cost; less waste
Savory and peppermint[82]	100–175 °C, 60–70 bar, 1.0 mL min^{-1}	SFE (CO_2) and hydrodistillation	Shorter extraction time; selective for oxygenates (no waxes or other plant products)
Oregano[83]	125 °C, 20 bar, 1.0 ml min^{-1}	hydrodistillation	Quicker; more efficient; cheaper
Ginkgo biloba[84a]	RT, 101 bar, 1.5–2.0 ml min^{-1}	Boiling ethanol, methanol, water and acetone	Good for thermally sensitive compounds
St John's wort[85a]	RT, 101 bar, static	Ultrasound (water and methanol)	
Lime peel[86]	130 °C, >20 bar, 1 ml min^{-1}	VOC extraction with sonication (CH_2Cl_2 and hexane) and hydrodistillation	Most selective for oxygenates; quicker
Morinda citrifolia[87]	220 °C, 70 bar, 4 ml min^{-1}	Not reported	None highlighted

[a]Pressurized water (not NCW).

Table 4.7. As with SFE, the plant residues are not contaminated with harmful VOCs and therefore the fibres and cellulose are available for further uses.

It should be noted that as with SFE, less harmful VOCs such as ethanol can be used to optimize the extraction of particular classes of compounds. Recently, NCW (100 °C, 14.8 bar, 10 mL min^{-1}) and pressurized 80% ethanol were used in the extraction of gypenosides from *Gynostemma pentaphyllum*. Water extracted 107 mg g^{-1} of desired biologically active components, whereas ethanol extracted 164 mg g^{-1}.[88] The compounds were unfortunately purified using traditional HPLC and a further solvent, methanol, was introduced to the process. The combined extraction and analysis of the compound using just NCW and ethanol would have been more desirable, especially as there are many benefits to using NCW as the mobile phase.

The use of NCW as the mobile phase in liquid chromatography was recently reviewed.[89] In this area, in addition to its green credentials, NCW is compatible with a range of detection methods; flame ionization detection, mass spectrometry (MS) and UV (to short wavelengths). The reason for the recent growth in this area is the development of more thermally stable stationary phases. It has been used to analyse a growing number of analytes (alkylbenzenes, phenols, ketones, carboxylic acids, amino acids, carbohydrates and some pharmaceuticals). For example, carbohydrates (monosaccharides, disaccharides and sugar alcohols) have been separated using a strong cation-exchange resin, which withstood the high temperatures well.[90] However, the separations were dependent on temperature and some sugars degraded at high temperatures. Therefore, when using NCW, the stability of the analyte rather than of the column material is probably the limiting factor in many separations. These separation and analytical techniques may well find application in the green separation and characterization of chemicals from biorefineries.

The technique can be coupled with NCW extraction methods.[91] Overall, this approach significantly decreases the amount of solvent used in an analytical laboratory. These extraction methods have also been coupled with capillary electrophoresis (CE) and CE–MS in the isolation, separation and characterization of antioxidant extracts from rosemary.[92,93]. NCW has also been used with enzyme catalysis to give a new, environmentally friendly method for analysing antioxidant content in onion waste.[94] The new method was quicker, higher yielding and used 100 times less organic solvents. Because of the high temperature of the extraction, enzymes from hyperthermophilic bacteria were used for this process. There is the potential that this technique developed for analytical chemistry could be modified and also carried out on a larger scale within a biorefinery. With the increasing importance of natural antioxidants, such methods will continue to have growing importance and impact.

4.2.2.2 Chemical Synthesis

Because the solubility of organic molecules in water increases as its temperature rises, chemical transformations performed in NCW are becoming more common.[80,95,96] Reactions performed include acid and base catalysis (*e.g.* hydrolysis reactions, Table 4.8) using the enhanced dissociation of NCW that eliminates the need for any added acid or base and subsequent neutralization and salt disposal. However, it should be noted that many of the effects that enhance reactivity in ambient water (*e.g.* hydrophobic effect) are less evident in this medium.

Hydrolysis of bio-sourced molecules has been investigated by several groups. Starch in bagasse, from previously extracted ginger root, could be rapidly

Table 4.8 Compounds that have been hydrolysed in NCW.[96]

Small molecules	Polymers
Ethers	Nylon
Esters	PET
Amides	Polycarbonate
Nitriles	Phenolic resin
Amines	Epoxy resin
Nitroalkane	Cellulose
Alkyl halide	Chitin
Glucose	Vegetable oil
Fructose	

hydrolysed in NCW at 300 °C to give high yields of reducing sugars.[97] Also, many hydrolysis reactions are more rapid in NCW than in other media. In the hydrolysis of β-pinene, 90% conversion was achieved in 20 min using water at 200 °C, whereas mixed alcohol–water mixtures took on average twice as long to achieve the same conversion.[98] Importantly, in control reactions using 100% ethanol no hydrolysis was observed, indicating that alcohols do not significantly contribute to the *in situ* acid catalyst formation. Unfortunately, in NCW extensive elimination and dehydration reactions occurred after the initial hydrolysis, ultimately yielding hydrocarbon products such as limonene. However, when a gas expanded phase was used (ethanol–water–carbon dioxide or acetone–water–carbon dioxide) the fraction of alcohols (terpineols) was increased. The use of NCW or carbon dioxide expanded media for *in situ* acid formation and catalysis has recently been reviewed.[99] In addition to hydrolysis reactions, NCW has also been used as a medium for condensation reactions that are conventionally acid or base catalysed,[100] including Claisen–Schmidt condensations and cross aldol reactions. Although the conversions and yields for these reactions were not exceptional, such processes show significant promise for NCW as a reaction medium. In some cases, the situation may be improved by adding a co-solvent. In a 50:50 water–ethanol mixture at high temperature and pressure, Poliakoff, Fraga-Dubreuil and co-workers have prepared phthalimide derivatives in high yields (Figure 4.12).[101] This procedure is normally performed in high boiling point solvents such as DMF or dioxane, and is therefore a significant green improvement for these widely used organic compounds.

More recently, reactions have been performed rapidly in NCW by using microwave reactors.[43,102] Some examples are shown in Figure 4.13. Although microwave compatible vessels capable of withstanding the high pressure and potentially corrosive nature of the NCW are required, these are commercially available and described in the papers by Leadbeater and Kappe. The commercial availability of the specialized instrumentation and the shortened reaction times are likely to increase activity in this area over the next 5 years.

A wide range of organic reactions have also been performed in SCW including hydrogenations, eliminations, condensations, hydrations and partial oxidations.[79] It has also been used in depolymerization reactions of natural and synthetic polymers. However, because SCW is more reactive than NCW,

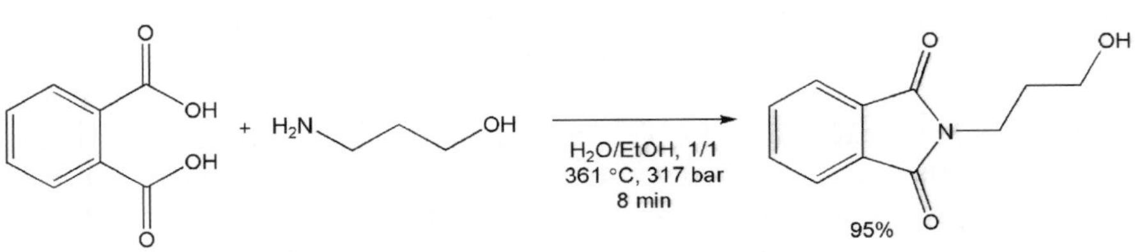

Figure 4.12   Phthalimide synthesis in a high pressure and temperature ethanol–water mixture.

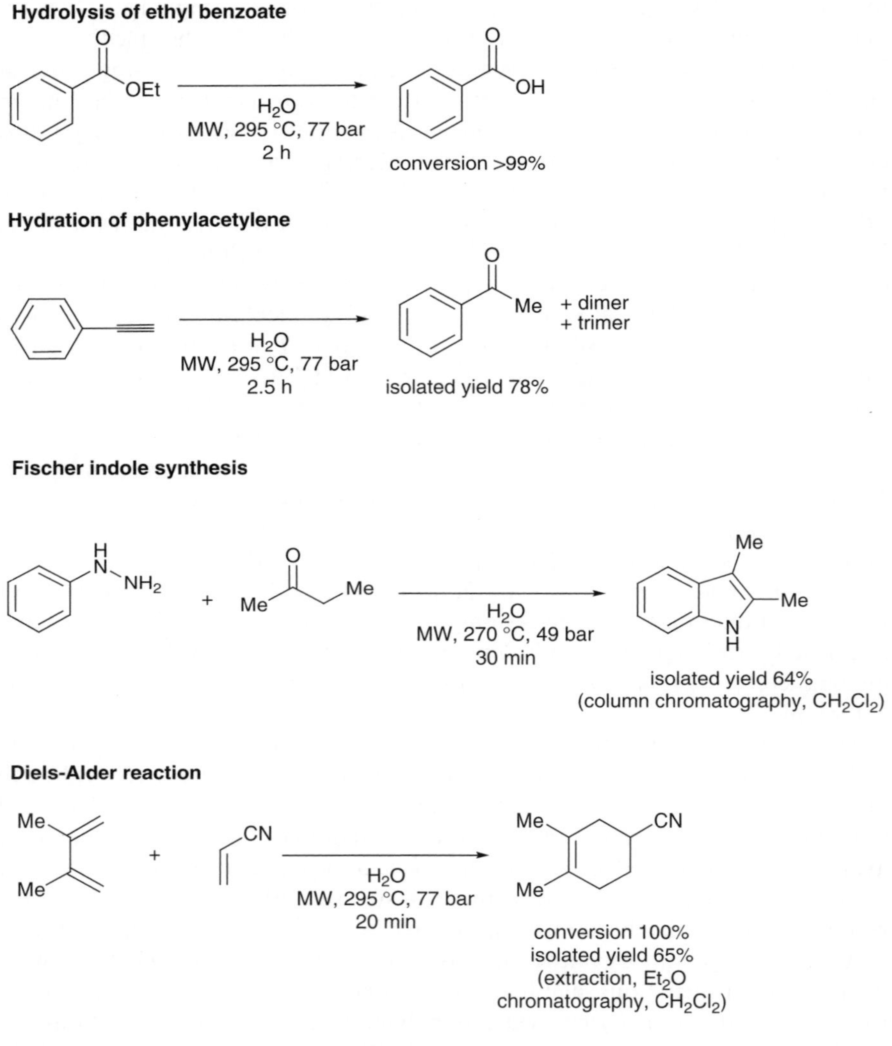

Hydrolysis of ethyl benzoate

Hydration of phenylacetylene

Fischer indole synthesis

Diels-Alder reaction

Figure 4.13 Some microwave assisted organic reactions in NCW.

special precautions have to be taken and problems can arise due to decomposition by products if the reaction conditions (temperature, pressure and time) are not controlled carefully. However, Poliakoff has pioneered the use of continuous-flow reactors in this field and this has led to significant advances. ε-Caprolactam, used in the manufacture of nylon-6, has been prepared in 90% yield through a continuous two-step hydrolysis and cyclization process in SCW.[103] No catalyst or additional solvent was needed and reaction times were dramatically reduced in comparison with other procedures that yield this important chemical. A variety of alkyl aromatics have been oxidized selectively

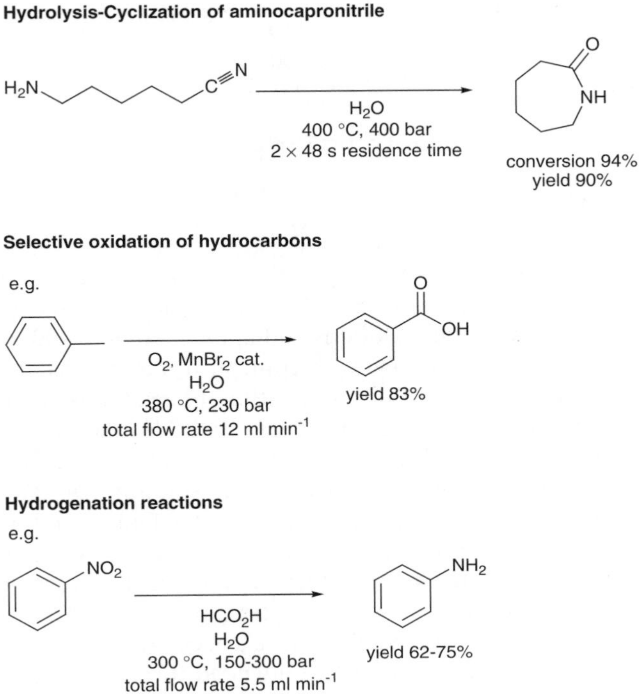

Hydrolysis-Cyclization of aminocapronitrile

H$_2$O
400 °C, 400 bar
2 × 48 s residence time

conversion 94%
yield 90%

Selective oxidation of hydrocarbons

e.g.

O$_2$, MnBr$_2$ cat.
H$_2$O
380 °C, 230 bar
total flow rate 12 ml min^{-1}

yield 83%

Hydrogenation reactions

e.g.

HCO$_2$H
H$_2$O
300 °C, 150-300 bar
total flow rate 5.5 ml min^{-1}

yield 62-75%

Figure 4.14 Continuous flow reactions in SCW.

to acids or aldehydes using a continuous-flow reactor (Figure 4.14).[104] In this study, hydrogen peroxide was thermally decomposed to yield oxygen, which was used as the oxidant. In another study, hydrogen was generated for reductions using the thermal decomposition of formic acid or formate salts.[105] Although this process is not intended for industrial use, in a research laboratory it is a convenient and safe way to perform hydrogenation reactions. Yields for these continuous hydrogenation reactions were comparable to those obtained using a batch reactor, but reaction times were significantly reduced (20–30 s *vs* 3 h).

4.2.2.3 Materials Synthesis

The use of supercritical fluids, including SCW and NCW, in inorganic materials synthesis and the preparation of nanoparticles was recently reviewed.[75,106] The hydrolysis and dehydration of metal nitrates and metal–organic precursors in supercritical water is also known as *hydrothermal synthesis* (Figure 4.15).

Recent examples include the batch synthesis of CoAl$_2$O$_4$ nanocrystals and the continuous synthesis of nano-hydroxyapatite.[107,108] In the first example, the researchers wanted to prepare metal oxide particles that could be easily

$$M(NO_3)_2 \xrightarrow[\substack{SCW \\ 375\text{-}400\,°C \\ 200\text{-}300\ bar}]{} MO_x$$

M = Ce, Cr, Pr, Fe etc.

Figure 4.15   Typical synthesis of metal oxides in SCW.

dispersed in non-aqueous solvents or polymers.[108] By taking advantage of the solubility of organic molecules in SCW, hexanoic acid or 1-hexylamine was added to $CoAl_2(OH)_5$ aqueous solutions and upon heating to 400 °C at 380 bar for 10 min, $CoAl_2O_4$ particles grew with controlled dimensions and were capped with the organic ligands. For the continuous synthesis of nano-hydroxyapatite, a continuous water feed at 400 °C and 240 bar was used. In this study, in addition to VOC solvents, no organic templating agents, ligands or expensive metal precursors were required. Basic solutions of calcium nitrate and ammonium phosphate were pumped to meet at a T-piece and then brought to meet a superheated water feed in a countercurrent reactor where the reaction occurred. In general, at high temperatures well-defined nanocrystalline rods formed whose size could be controlled by temperature. Therefore, given the biocompatible nature of water, SCW is an excellent method of synthesis for materials such as hydroxyapatite that are intended for biological applications.

4.2.2.4   Supercritical Water Oxidation (SCWO)

SCW is an excellent medium for the total oxidation of unwanted and hazardous organic compounds such as those that need to be removed from wastewater and process streams (Table 4.9).[79,109,110] NCW and $scCO_2$ are also being investigated as green alternatives in the design of environmentally friendly processes for pollutant recovery and recycling (including soil remediation and nuclear reprocessing). Supercritical water oxidation (SCWO) can rapidly transform 99.9999% of contaminants, at much lower temperatures than incineration in air.

In less than 1 min of residence time, organic carbon is converted to carbon dioxide and importantly, nitrogenous compounds are converted to gaseous nitrogen and not to polluting NO_x. Halogens are converted to HX and sulfur compounds to sulfuric acid (no SO_x). The typical output of a SCWO plant contains carbon dioxide, nitrogen, water, hydrochloric acid, sulfuric acid, phosphoric acid and trace amounts of acetic acid and nitrous oxide. The corrosive nature of the SCW and also some of the acidic products are the major challenges in this area, and many plants built during the late 1990s have already closed as a result of corrosion and plugging problems.[109] However, there is hope that new reactor designs (*e.g.* the transpiring wall reactor) will help to overcome these problems. Further details on this process and reactor designs can be found in the 2006 review by Bermejo and Cocero.[109]

Table 4.9 Advantages of SCWO for waste treatment.

On-site treatment
Complete destruction of organic waste
Totally enclosed process and >50% of available heat easily recovered
Competitive unit cost
No major permitting issues
Good public acceptability

SCWO is also being investigated as a technique for biomass processing, as an alternative to fermentation processes. A view cell was recently used to look at the decomposition of wood under different conditions including temperature, pressure and oxygen concentration, in order to gain a better understanding of SCWO and its potential in this area.[111]

4.3 Summary and Outlook for the Future

As with many other alternative solvents, it is probably the mindset of the chemist that inhibits the more extensive adoption of SCF technologies. The main drawback is a significant initial investment, and therefore more extensive collaborations in this field between industrial organizations, academia, chemists and chemical engineers should be encouraged. However, despite these challenges to progress, large scale apparatus has been used for extraction for many years and new areas (dry cleaning and polymer synthesis) have also adopted this technology outside the laboratory (Chapter 10). Attention must be paid to the economic viability of using scCO$_2$ as a solvent; the green benefits of scCO$_2$ will be realized only if these processes cost less than their conventional analogues.[11] There is a significant energy implication for working at high pressures. To some extent this is overcome through chemical engineering, but it can also be overcome by using carbon dioxide expanded media (Chapter 9). Beckman[11] put forward the following rules for operating a carbon dioxide based process economically:

Operate at high concentrations
Operate at the lowest possible pressure
Recover products without high-pressure drops
Operate the process continuously
Recover and reuse homogeneous catalysts and CO$_2$-philes.

Clearly, this is achievable, as areas where scCO$_2$ is currently being used include dyeing and cleaning of fibres and textiles (Micell, USA), polymerization and polymer processing (DuPont and Xerox, USA), extraction of natural products (Botanix Ltd, UK) and catalysis (Thomas Swan & Co., UK). A substantial amount of current research is focused on using scCO$_2$ as a reaction medium for chemical synthesis. Only a selected few reactions could be discussed here and many more examples can be found in the reviews and journal articles referenced in this chapter and elsewhere. In addition to this, related areas such as liquid carbon dioxide and carbon dioxide expanded solvents should not be overlooked. Many additives and complex modifiers are being used to facilitate reactions in $scCO_2$ and perhaps the use of a small amount of VOC (perhaps from a bio-feedstock) could be justified in order to reduce the cost of a process and therefore lead to its uptake by industry. In addition to this, continued research into biphasic systems, such as carbon dioxide–water, carbon dioxide–ionic liquids, carbon dioxide–PEG–surfactants and carbon dioxide–solids (including hetero-geneous catalysts), is needed to deliver pure products and reduced cost to future end users of this technology.

In the field of high-pressure water, NCW has already provided very pro-mising results in the field of analytical chemistry for the extraction and separation of natural products. Research in this area is likely to increase as the technique becomes more widely available and recognized. Additionally, NCW has shown itself to be a versatile solvent in organic synthesis, particularly where acids or bases are normally used to catalyse a reaction. With the introduction of microwave instruments for NCW work, research in this area is likely to con-tinue to grow. Unfortunately, large scale SCWO plants have suffered from engineering problems. Therefore, further research needs to be performed in this area. However, on a laboratory scale at least, the use of continuous-flow reactors has given excellent results for a range of synthetic transformations of small organic molecules and the preparation of high-value materials. Appa-ratus for such techniques is currently put together in-house by researchers; the more widespread uptake of continuous-flow SCW techniques is dependent on commercial availability of pre-assembled equipment. In summary, although the use of NCW and SCW is less advanced than that of $scCO_2$, they show excep-tional promise as green solvents for a wide range of applications.

References

1. M. A. McHugh and V. J. Krukonis, *Supercritical Fluid Extraction: Principles and Practice*, Butterworth-Heinemann, Boston, 1994.
2. A. A. Clifford, *Fundamentals of Supercritical Fluids*, Oxford University Press, Oxford, 1998.
3. Y. Arai, T. Sako and Y. Takebayashi, *in Supercritical Fluids: Molecular Interactions, Physical Properties and New Applications*, Springer, Berlin, 2002.
4. S. Howdle, in http://www.nottingham.ac.uk/~pczctg/Video_Clip_5.htm, accessed 2008.
5. R. M. Oag, P. J. King, C. J. Mellor, M. W. George, J. Ke and M. Poliakoff, *Anal. Chem.*, 2003, **75**, 479.
6. P. E. Savage, S. Gopalan, T. I. Mizan, C. J. Martino and E. E. Brock, *AIChE J.*, 1995, **41**, 1723.
7. P. G. Jessop, Y. Hsiao, T. Ikariya and R. Noyori, *J. Am. Chem. Soc.*, 1996, **118**, 344.

8. J. L. Kendall, D. A. Canelas, J. L. Young and J. M. DeSimone, *Chem. Rev.*, 1999, **99**, 543.
9. A. Furstner, L. Ackermann, K. Beck, H. Hori, D. Koch, K. Langemann, M. Liebl, C. Six and W. Leitner, *J. Am. Chem. Soc.*, 2001, **123**, 9000.
10. K. Wittmann, W. Wisniewski, R. Mynott, W. Leitner, C. L. Kranemann, T. Rische, P. Eilbracht, S. Kluwer, J. M. Ernsting and C. L. Elsevier, *Chem. Eur. J.*, 2001, **7**, 4584.
11. E. J. Beckman, *Environ. Sci. Technol.*, 2002, **36**, 347A.
12. J. F. Brennecke and J. E. Chateauneuf, *Chem. Rev.*, 1999, **99**, 433.
13. T. Sarbu, T. Styranec and E. J. Beckman, *Nature*, 2000, **405**, 165.
14. V. K. Potluri, J. H. Xu, R. Enick, E. Beckman and A. D. Hamilton, *Org. Lett.*, 2002, **4**, 2333.
15. B. S. Murray, E. Dickinson, D. A. Clarke and C. M. Rayner, *Chem. Commun.*, 2006, 1410.
16. P. G. Jessop, T. Ikariya and R. Noyori, *Chem. Rev.*, 1999, **99**, 475.
17. S. M. Pourmortazavi and S. S. Hajimirsadeghi, *J. Chromatogr. A*, 2007, **1163**, 2.
18. E. Reverchon and I. De Marco, *J. Supercrit. Fluids*, 2006, **38**, 146.
19. R. M. Smith, *J. Chromatogr. A*, 1999, **856**, 83.
20. L. C. McKenzie, J. E. Thompson, R. Sullivan and J. E. Hutchison, *Green Chem.*, 2004, **6**, 355.
21. N. Asfaw, P. Licence, A. A. Novitskii and M. Poliakoff, *Green Chem.*, 2005, **7**, 352.
22. F. E. I. Deswarte, J. H. Clark, J. J. E. Hardy and P. M. Rose, *Green Chem.*, 2006, **8**, 39.
23. J. H. Clark, V. Budarin, F. E. I. Deswarte, J. J. E. Hardy, F. M. Kerton, A. J. Hunt, R. Luque, D. J. Macquarrie, K. Milkowski, A. Rodriguez, O. Samuel, S. J. Tavener, R. J. White and A. J. Wilson, *Green Chem.*, 2006, **8**, 853.
24. J. M. DeSimone and W. Tumas, *in Green Chemistry Using Liquid and Supercritical Carbon Dioxide*, Oxford University Press Oxford, 2003.
25. E. J. Beckman, *J. Supercrit. Fluids*, 2004, **28**, 121.
26. R. S. Oakes, A. A. Clifford and C. M. Rayner, *J. Chem. Soc., Perkin Trans. 1*, 2001, 917.
27. P. G. Jessop and W. Leitner, *in Chemical Synthesis using Supercritical Fluids*, VCH, Weinheim, 1999.
28. A. A. Clifford, K. Pople, W. J. Gaskill, K. D. Bartle and C. M. Rayner, *Chem. Commun.*, 1997, 595.
29. A. A. Clifford, K. Pople, W. J. Gaskill, K. D. Bartle and C. M. Rayner, *J. Chem. Soc., Faraday Trans.*, 1998, **94**, 1451.
30. R. S. Oakes, A. A. Clifford, K. D. Bartle, M. T. Petti and C. M. Rayner, *Chem. Commun.*, 1999, 247.
31. F. M. Kerton, G. A. Lawless and S. P. Armes, *J. Mater. Chem.*, 1997, **7**, 1965.
32. P. G. Jessop, *J. Supercrit. Fluids*, 2006, **38**, 211.
33. P. G. Jessop, T. Ikariya and R. Noyori, *Nature*, 1994, **368**, 231.

34. M. J. Burk, S. Feng, M. F. Gross and W. Tumas, *J. Am. Chem. Soc.*, 1995, **117**, 8277.
35. P. G. Jessop, T. Ikariya and R. Noyori, *Organometallics*, 1995, **14**, 1510.
36. M. F. Sellin, I. Bach, J. M. Webster, F. Montilla, V. Rosa, T. Aviles, M. Poliakoff and D. J. Cole-Hamilton, *J. Chem. Soc., Dalton Trans*, 2002, 4569.
37. I. Bach and D. J. Cole-Hamilton, *Chem. Commun.*, 1998, 1463.
38. S. Fujita, S. Fujisawa, B. M. Bhanage, Y. Ikushima and M. Arai, *Eur. J. Org. Chem.*, 2004, 2881.
39. Y. L. Hu, W. P. Chen, A. M. B. Osuna, A. M. Stuart, E. G. Hope and J. L. Xiao, *Chem. Commun.*, 2001, 725.
40. M. A. Carroll and A. B. Holmes, *Chem. Commun.*, 1998, 1395.
41. D. K. Morita, D. R. Pesiri, S. A. David, W. H. Glaze and W. Tumas, *Chem. Commun.*, 1998, 1397.
42. N. Shezad, R. S. Oakes, A. A. Clifford and C. M. Rayner, *Tetrahedron Lett.*, 1999, **40**, 2221.
43. N. E. Leadbeater, *Chem. Commun.*, 2005, 2881.
44. N. Shezad, A. A. Clifford and C. M. Rayner, *Green Chem.*, 2002, **4**, 64.
45. C. J. Smith, M. W. S. Tsang, A. B. Holmes, R. L. Danheiser and J. W. Tester, *Org. Biomol. Chem.*, 2005, **3**, 3767.
46. P. Licence, J. Ke, M. Sokolova, S. K. Ross and M. Poliakoff, *Green Chem.*, 2003, **5**, 99.
47. P. Clark, M. Poliakoff and A. Wells, *Adv. Synth. Catal.*, 2007, **349**, 2655.
48. J. R. Hyde, P. Licence, D. Carter and M. Poliakoff, *Appl. Catal. A*, 2001, **222**, 119.
49. R. Ciriminna, M. L. Carraro, S. Campestrini and M. Pagliaro, *Adv. Synth. Catal.*, 2008, **350**, 221.
50. R. A. Bourne, J. G. Stevens, J. Ke and M. Poliakoff, *Chem. Commun.*, 2007, 4632.
51. P. Licence, W. K. Gray, M. Sokolova and W. K. Gray, *J. Am. Chem. Soc.*, 2005, **127**, 293.
52. D. Hancu, J. Green and E. J. Beckman, *Acc. Chem. Res.*, 2002, **35**, 757.
53. D. Hancu, H. Green and E. J. Beckman, *Ind. Eng. Chem. Res.*, 2002, **41**, 4466.
54. E. J. Beckman, *Green Chem.*, 2003, **5**, 332.
55. B. M. Bhanage, Y. Ikushima, M. Shirai and M. Arai, *Chem. Commun.*, 1999, 1277.
56. K. Burgemeister, G. Francio, H. Hugl and W. Leitner, *Chem. Commun.*, 2005, 6026.
57. K. Burgemeister, G. Francio, V. H. Gego, L. Greiner, H. Hugl and W. Leitner, *Chem. Eur. J.*, 2007, **13**, 2798.
58. S. Keskin, D. Kayrak-Talay, U. Akman and O. Hortacsu, *J. Supercrit. Fluids*, 2007, **43**, 150.
59. F. C. Liu, M. B. Abrams, R. T. Baker and W. Tumas, *Chem. Commun.*, 2001, 433.

60. Z. F. Zhang, E. Xie, W. J. Li, S. Q. Hu, J. L. Song, T. Jiang and B. X. Han, *Angew. Chem. Int. Ed.*, 2008, **47**, 1127.
61. R. A. Brown, P. Pollet, E. McKoon, C. A. Eckert, C. L. Liotta and P. G. Jessop, *J. Am. Chem. Soc.*, 2001, **123**, 1254.
62. M. F. Sellin, P. B. Webb and D. J. Cole-Hamilton, *Chem. Commun.*, 2001, 781.
63. P. B. Webb, M. F. Sellin, T. E. Kunene, S. Williamson, A. M. Z. Slawin and D. J. Cole-Hamilton, *J. Am. Chem. Soc.*, 2003, **125**, 15577.
64. M. T. Reetz, W. Wiesenhofer, G. Francio and W. Leitner, *Chem. Commun.*, 2002, 992.
65. H. R. Hobbs and N. R. Thomas, *Chem. Rev.*, 2007, **107**, 2786.
66. A. J. Mesiano, E. J. Beckman and A. J. Russell, *Chem. Rev.*, 1999, **99**, 623.
67. J. M. DeSimone, Z. Guan and C. S. Elsbernd, *Science*, 1992, **257**, 945.
68. C. D. Wood, A. I. Cooper and J. M. DeSimone, *Curr. Opin. Solid State Mater. Sci.*, 2004, **8**, 325.
69. J. M. DeSimone, E. E. Maury, Y. Z. Menceloglu, J. B. McClain, T. J. Romack and J. R. Combes, *Science*, 1994, **265**, 356.
70. A. I. Cooper, *J. Mater. Chem.*, 2000, **10**, 207.
71. A. I. Cooper, *Adv. Mater.*, 2001, **13**, 1111.
72. A. I. Cooper, *Adv. Mater.*, 2003, **15**, 1049.
73. T. Hasell, K. J. Thurecht, R. D. W. Jones, P. D. Brown and S. M. Howdle, *Chem. Commun.*, 2007, 3933.
74. S. Moisan, V. Martinez, P. Weisbecker, F. Cansell, S. Mecking and C. Aymonier, *J. Am. Chem. Soc.*, 2007, **129**, 10602.
75. E. Reverchon and R. Adami, *J. Supercrit. Fluids*, 2006, **37**, 1.
76. C. A. Bessel, G. M. Denison, J. M. DeSimone, J. DeYoung, S. Gross, C. K. Schauer and P. M. Visintin, *J. Am. Chem. Soc.*, 2003, **125**, 4980.
77. R. A. Pai, R. Humayun, M. T. Schulberg, A. Sengupta, J. N. Sun and J. J. Watkins, *Science*, 2004, **303**, 507.
78. J. W. Wang, Y. D. Xia, W. X. Wang, M. Poliakoff and R. Mokaya, *J. Mater. Chem.*, 2006, **16**, 1751.
79. P. E. Savage, *Chem. Rev.*, 1999, **99**, 603.
80. A. R. Katritzky, D. A. Nichols, M. Siskin, R. Murugan and M. Balasubramanian, *Chem. Rev.*, 2001, **101**, 837.
81. V. Fernandez-Perez, M. M. Jimenez-Carmona and M. D. L. de Castro, *Analyst*, 2000, **125**, 481.
82. A. Kubatova, A. J. M. Lagadec, D. J. Miller and S. B. Hawthorne, *Flavour Frag. J.*, 2001, **16**, 64.
83. R. S. Ayala and M. D. L. de Castro, *Food Chem.*, 2001, **75**, 109.
84. Q. Y. Lang and C. M. Wai, *Green Chem.*, 2003, **5**, 415.
85. M. Mannila and C. M. Wai, *Green Chem.*, 2003, **5**, 387.
86. O. Chienthavorn and W. Insuan, *Anal. Lett.*, 2004, **37**, 2393.
87. A. Shotipruk, J. Kiatsongserm, P. Pavasant, M. Goto and M. Sasaki, *Biotechnol. Progr.*, 2004, **20**, 1872.
88. C. H. Chen, T. Y. Huang, M. R. Lee, S. L. Hsu and C. M. J. Chang, *Ind. Eng. Chem. Res.*, 2007, **46**, 8138.

89. R. M. Smith, *J. Chromatogr. A*, 2008, **1184**, 441.
90. J. Tiihonen, E. L. Peuha, M. Latva-Kokko, S. Silander and E. Paatero, *Sep. Purif. Technol.*, 2005, **44**, 166.
91. R. Tajuddin and R. M. Smith, *Analyst*, 2002, **127**, 883.
92. M. Herrero, D. Arraez-Roman, A. Segura, E. Kenndler, B. Gius, M. A. Raggi, E. Ibanez and A. Cifuentes, *J. Chromatogr. A*, 2005, **1084**, 54.
93. A. Crego, E. Ibanez, E. Garcia, R. R. de Pablos, F. J. Senorans, G. Reglero and A. Cifuentes, *Eur. Food Res. Technol.*, 2004, **219**, 549.
94. C. Turner, P. Turner, G. Jacobson, K. Almgren, M. Waldeback, P. Sjoberg, E. N. Karlsson and K. E. Markides, *Green Chem.*, 2006, **8**, 949.
95. M. Siskin and A. R. Katritzky, *Chem. Rev.*, 2001, **101**, 825.
96. N. Akiya and P. E. Savage, *Chem. Rev.*, 2002, **102**, 2725.
97. S. R. M. Moreschi, A. J. Petenate and M. A. A. Meireles, *J. Agric. Food Chem.*, 2004, **52**, 1753.
98. T. S. Chamblee, R. R. Weikel, S. A. Nolen, C. L. Liotta and C. A. Eckert, *Green Chem.*, 2004, **6**, 382.
99. J. P. Hallett, P. Pollet, C. L. Liotta and C. A. Eckert, *Acc. Chem. Res.*, 2008, **41**, 458.
100. S. A. Nolen, C. L. Liotta, C. A. Eckert and R. Glaser, *Green Chem.*, 2003, **5**, 663.
101. J. Fraga-Dubreuil, G. Comak, A. W. Taylor and M. Poliakoff, *Green Chem.*, 2007, **9**, 1067.
102. J. M. Kremsner and C. O. Kappe, *Eur. J. Org. Chem.*, 2005, 3672.
103. C. Yan, J. Fraga-Dubreuil, E. Garcia-Verdugo, P. A. Hamley, M. Poliakoff, I. Pearson and A. S. Coote, *Green Chem.*, 2008, **10**, 98.
104. J. Fraga-Dubreuil, E. Garcia-Verdugo, P. A. Hamley, E. M. Vaquero, L. M. Dudd, I. Pearson, D. Housley, W. Partenheimer, W. B. Thomas, K. Whiston and M. Poliakoff, *Green Chem.*, 2007, **9**, 1238.
105. E. Garcia-Verdugo, Z. M. Liu, E. Ramirez, J. Garcia-Serna, J. Fraga-Dubreuil, J. R. Hyde, P. A. Hamley and M. Poliakoff, *Green Chem.*, 2006, **8**, 359.
106. C. Aymonier, A. Loppinet-Serani, H. Reveron, Y. Garrabos and F. Cansell, *J. Supercrit. Fluids*, 2006, **38**, 242.
107. A. A. Chaudhry, S. Haque, S. Kellici, P. Boldrin, I. Rehman, A. K. Fazal and J. A. Darr, *Chem. Commun.*, 2006, 2286.
108. D. Rangappa, S. Ohara, T. Naka, A. Kondo, M. Ishii and T. Adschiri, *J. Mater. Chem.*, 2007, **17**, 4426.
109. M. D. Bermejo and M. J. Cocero, *AIChE J.*, 2006, **52**, 3933.
110. H. Schmieder and J. Abeln, *Chem. Eng. Technol.*, 1999, **22**, 903.
111. D. Shoji, N. Kuramochi, K. Yui, H. Uchida, K. Itatani and S. Koda, *Ind. Eng. Chem. Res.*, 2006, **45**, 5885.

CHAPTER 5

Renewable Solvents

5.1 Introduction

Many solvents can be obtained from renewable feedstocks and can be used as 'slot-in' alternatives for current VOCs without any need for modification of equipment or procedure. Because of the large number of oxygens in biomass-sourced materials such as cellulose and starch, it is not surprising that most renewable solvents have oxygen-containing functional groups, alcohols, esters and ethers being the most common. However, many currently employed solvents also contain these groups.[1] The most extensively used group of VOC solvents that cannot be bio-sourced are chlorinated hydrocarbons such as methylene chloride. However, blends of bio-solvents can be made and used in many applications where these are normally used. Hydrocarbons including aromatics could potentially be bio-sourced through transformations of cellulose and lignocellulose.

Bio-solvents are produced through a *biorefinery* approach to commodity chemicals manufacture. A biorefinery can be defined as a facility that integrates biomass conversion processes and equipment to produce fuels, power, and chemicals from biomass. Research in this area and its interface with green chemistry has expanded dramatically in the last 2 years.[2–5] Just as today petroleum refineries produce multiple fuels and products from oil, industrial biorefineries should be able to produce many of these products in the future. The U.S. Department of Energy has identified 12 bio-sourced platform chemicals or building blocks that can be produced either biologically or chemically from natural carbohydrate feedstocks (Figure 5.1). Because of the many acid- and alcohol-functionalized molecules in this group, significant research is ongoing in the field of polymer chemistry to yield new bio-derived polyesters using esterification reactions. These include Sorona produced by DuPont and CORTERRA PTT produced by Shell: both use 1,3-propanediol as a feedstock.

RSC Green Chemistry Book Series
Alternative Solvents for Green Chemistry
By Francesca M. Kerton
© Francesca M. Kerton 2009
Published by the Royal Society of Chemistry, www.rsc.org

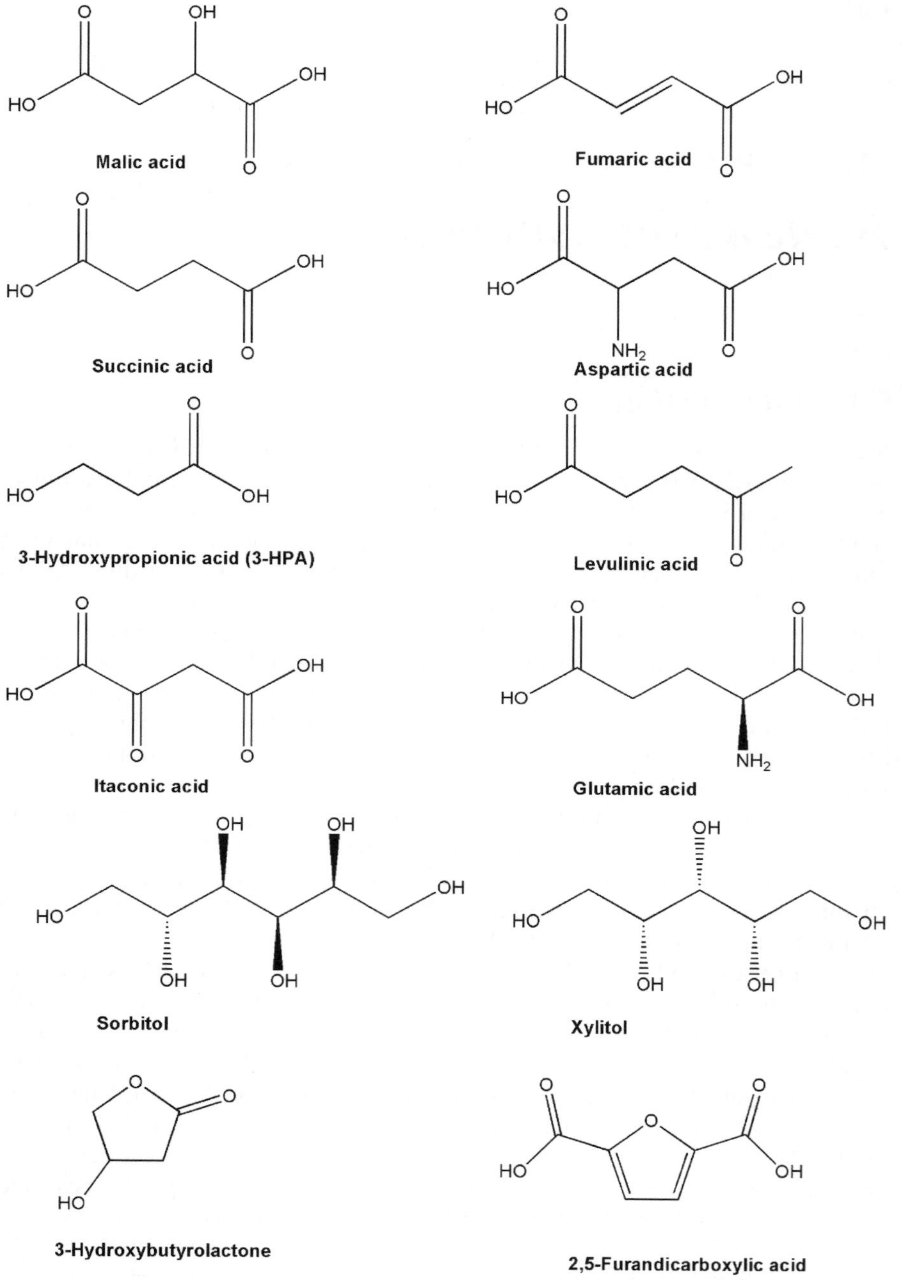

Figure 5.1   Bio-sourced platform chemicals.

Table 5.1   Biomass feedstocks.

Waste materials	Agricultural, wood, and urban wastes, crop residue
Forest products	Wood, logging residues, trees, shrubs
Energy crops	Starch crops such as corn, wheat, and barley, sugar crops, grasses, vegetable oils, hydrocarbon plants (*e.g. Pittosporum resiniferum, Euphorbia lathyris*)
Aquatic biomass	Algae, waterweed (including seaweed), water hyacinth

Another approach to biomass-derived chemical production is the *two-platform concept* where the production of syngas (synthesis gas) from biomass gasification, or other technologies, is used to produced methanol or hydrocarbons through Fischer–Tropsch technology.[5]

The variety of feedstocks to generate platform chemicals or fuels from biomass is shown in Table 5.1.[4] They can also be divided into three groups according to their chemical make-up: cellulosic biomass, starch- and sugar-derived biomass (or edible biomass) and triglyceride based biomass. The cost of these feedstocks depends on regional issues and market forces. However, they generally increase in price in the order: cellulosic biomass, starch (and sugar) based biomass, triglyceride based biomass.[4] Unfortunately, there is growing controversy surrounding the use of edible biomass because of current food shortages in some countries and increasing food costs globally. These may or may not be the result of the booming biofuel business. The cost of converting the biomass into chemicals including fuels is cheapest for triglycerides and most expensive for cellulosic materials. However, extensive research is ongoing in the area of cellulose conversion, and cellulose-derived chemicals and fuels have a promising future. In 2008, General Motors announced a partnership with Coskata, Inc. to produce cellulosic ethanol cheaply, with an eventual goal of $1 per U.S. gallon ($0.30 L^{-1}) for the fuel. The partnership plans to begin producing the fuel in large quantity by the end of 2008. By 2011 a full scale plant will come on line, capable of producing 50–100 million gallons of ethanol a year.[6]

The cost of crude oil has dramatically increased during the last decade (Figure 5.2), and bio-derived fuel production—bio-ethanol and biodiesel—has therefore increased significantly. These liquids can also be used as solvents in chemistry. Biodiesel production has also led to vast amounts of glycerol entering the market which can be used directly as a solvent (see below), or converted into diols, esters, ethers and a myriad of other chemicals.[7]

Although bio-sourced solvents are nominally green in terms of a life-cycle analysis, they are not perfect. They are still VOCs and have associated risks including atmospheric pollution, flammability and user exposure. Also, as is regularly highlighted in media coverage of biofuels, bio-sourced chemicals may not be carbon neutral because fertilizers and a significant amount of energy are used in their production. Therefore, in many cases it would be advisable to undertake a complete environmental economic analysis to assess the triple bottom line of social, economic and environmental advantages and disadvantages. Additionally, in most cases, as can be seen from their molecular

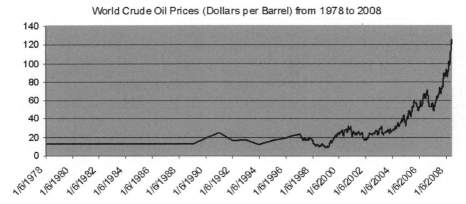

Figure 5.2 Crude oil prices over the last 30 years. Data from Energy Information Administration, US Government, http://www.eia.doe.gov/, web accessed June 2008.

structure (Figure 5.3), bio-solvents are not inert when compared to conventional solvents such as methylene chloride and toluene. For example, alcohols can undergo substitution, oxidation and dehydration reactions. Bio-sourced acetic acid can also be used as a solvent but is not discussed here because of its corrosive nature.

5.2   Chemical Examples

5.2.1   Alcohols including Glycerol

Ethanol is generally produced through fermentation of starch crops but routes from cellulose, which can come from waste materials, are gaining momentum. Methanol can be produced from syngas that can be obtained through biomass gasification. Ethanol and methanol are commonplace solvents in laboratories worldwide, but today are less widely used in reactions and separations than petroleum-sourced alternatives such as halogenated and aromatic solvents. Methanol and ethanol are both volatile, with low flash points and large explosion ranges (Table 5.2), which means there are significant hazards in their use especially when compared with many other alternative solvents including glycerol.

Ethanol is commonly used as a solvent of substances intended for human contact or consumption, including scents, flavourings, colourings, and medicines. It is widely used in the food industry and in the extraction of natural products, and is also used in thermometers. The physical properties of ethanol stem primarily from the presence of its hydroxyl group and the shortness of its carbon chain. The hydroxyl group is able to participate in hydrogen bonding, rendering ethanol more viscous and less volatile than less polar organic

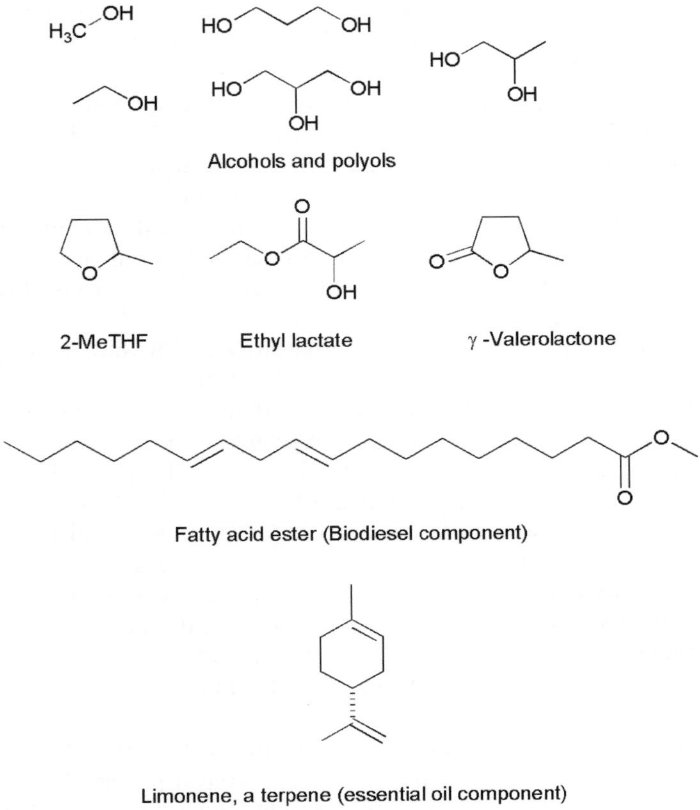

Alcohols and polyols

2-MeTHF   Ethyl lactate   γ-Valerolactone

Fatty acid ester (Biodiesel component)

Limonene, a terpene (essential oil component)

Figure 5.3 Some solvents available from renewable feedstocks.

Table 5.2 Comparison of solvent properties of methanol, ethanol and glycerol.

Property	MeOH	EtOH	Glycerol
Dielectric constant	32.66	24.3	42.5
Density/g cm^{-3}	0.79	0.79	1.26
Boiling point/°C	64.7	78.4	290
Melting point/°C	−97	−114	18
Viscosity/cP	0.6	1.2	629
pKa	15.5	15.9	14.4
Flash point/°C	12	16	160
Explosion range, lower/ upper limit/vol%	6.0/36	3.3/19	Not applicable
Vapour pressure/mm Hg at 20 °C	97	44	<1
Hildebrand/(Mpa)$^{1/2}$	29.7	26.2	36.2
Donor number/kcal mol^{-1}	19	31.5	—

compounds of similar molecular weight. It is a versatile solvent, miscible with water and with many organic solvents, including acetone, diethyl ether, glycerol, and toluene. It is also miscible with light aliphatic hydrocarbons, such as pentane and hexane. Its miscibility with water contrasts with that of longer chain alcohols (five or more carbon atoms), whose water miscibility decreases sharply as the number of carbons increases. The polar nature of the hydroxyl group means that ethanol is able to dissolve many ionic compounds, including sodium and potassium hydroxides and ammonium chloride and bromide. Because the ethanol molecule also has a non-polar end, it will also dissolve non-polar substances, including many essential oils and numerous flavourings, colourings and medicinal agents.

Methanol has similar physical properties to ethanol, but it is toxic and ethanol is therefore the preferred solvent in most applications, *e.g.* medicinal agents. However, in synthetic procedures methanol is more commonly used because of its greater volatility and ease of removal under vacuum.

As ethanol and methanol are common laboratory solvents, their application in extraction and reaction chemistry is not be discussed at length here; details on many procedures using these solvents can be found in chemistry textbooks and the primary literature. However, exciting new procedures using acid catalysis in aqueous ethanol for the esterification of platform molecules have recently been reported.[8,9] This reaction also highlights the reactivity of alcohols, as ethanol is one of the substrates in the reaction (Figure 5.4). It is likely that ethanol and water will continue to play a prominent role as solvents in the new transformation chemistries being developed.

Glycerol, which is a by-product of biodiesel production and other processes, is non-toxic and has promising physical and chemical properties as an alternative solvent.[10,11] It has a very high boiling point and negligible vapour pressure (Table 5.2), and can dissolve many organic and inorganic compounds. It is poorly miscible with water and some ethers and hydrocarbons. Therefore, in addition to distilling products from this solvent, simple extractions with solvents such as ether and ethyl acetate are also possible. It should also be noted that glycerol can be converted to methanol, ethanol, 1-propanol and propanediols through hydrogenolysis reactions, and is therefore a potential feedstock for other solvents.[2]

High conversions and selectivities have been obtained for a range of catalytic and stoichiometric reactions performed in glycerol including nucleophilic substitutions, stoichiometric ($NaBH_4$) and catalytic (H_2 with Pd–C) reductions, Heck and Suzuki couplings, and enzymatic transesterifications.[10,12] Although in most cases glycerol could not be described as the optimum alternative solvent, these studies demonstrate the potential that it holds for future investigations. In contrast, for yeast catalysed reductions of prochiral β-keto esters and ketones (Figure 5.5), excellent yields and selectivities were obtained.[13] Isolated yields and enantioselectivities were comparable with reactions in water, and superior to results obtained in ionic liquid or fluorous media. However, a significantly longer reaction time was needed to obtain the same conversions in glycerol compared with water. It should be noted that in

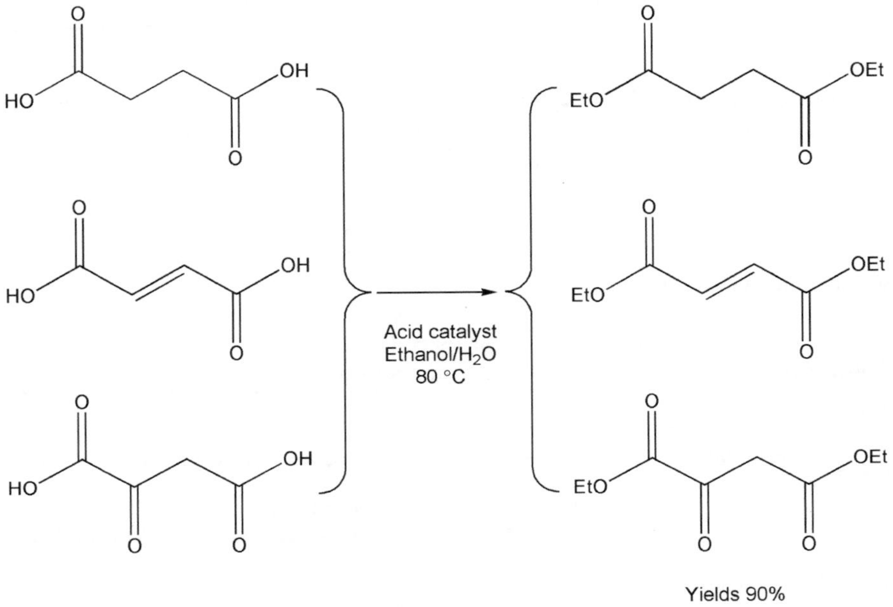

Figure 5.4 Esterification of diacid platform molecules in aqueous ethanol.

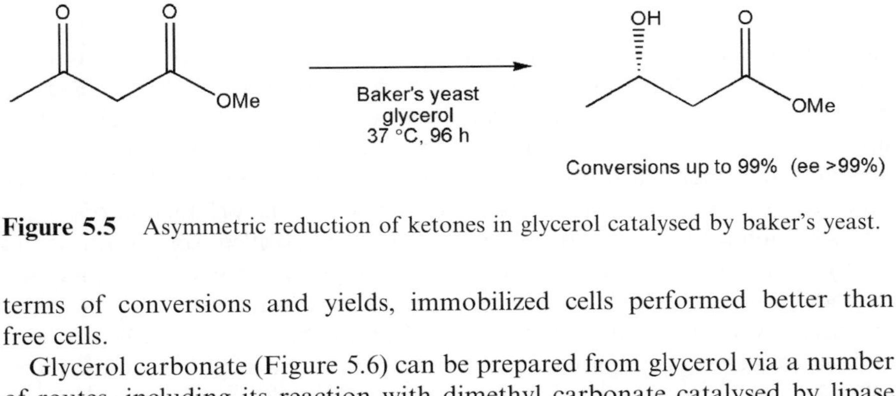

Figure 5.5 Asymmetric reduction of ketones in glycerol catalysed by baker's yeast.

terms of conversions and yields, immobilized cells performed better than free cells.

Glycerol carbonate (Figure 5.6) can be prepared from glycerol via a number of routes, including its reaction with dimethyl carbonate catalysed by lipase enzymes.[7,14] It has potential as a biosolvent for coatings, cosmetics and pharmaceuticals, and as a lubricant. However, as it is a relatively new material in the chemical industry, limited data are currently available.

5.2.2 Esters

Argonne National Laboratory received a 1998 Presidential Green Chemistry Challenge Award for the development of a novel membrane based process for producing lactate esters.[15] The process uses pervaporation membranes and

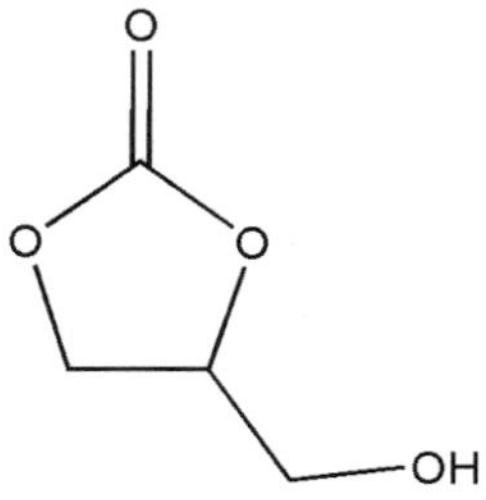

Figure 5.6   Glycerol carbonate.

Table 5.3   Industrial uses of esteric green solvents.[21]

Solvent	Industrial use
Glycerol carbonate	Non-reactive diluent in epoxy or polyurethane systems
Ethyl lactate	Degreaser
	Photo-resist carrier solvent
	Clean-up solvent in microelectronics and semiconductor manufacture
2-Ethylhexyl lactate	Degreaser
	Agrochemical formulations
Fatty acid esters	Biodegradable carrier oil for green inks
(and related compounds)	Coalescent for decorative paint systems
	Agrochemical/pesticide formulations

catalysts to dramatically reduce the required energy input and the amount of waste produced. Ammonium lactate, which is produced in a fermentation process, is thermally and catalytically cracked to produce lactic acid, which upon addition of an alcohol generates the ester. The ammonia and water by-products are separated through a selective membrane and recycled. This process, which uses carbohydrate feedstocks, has made the production of lactate esters economically competitive. In turn, due to the excellent solvent properties of ethyl lactate, it has become widely available as a bio-sourced and bio-degradable cleaning fluid (Table 5.3).[16] It has also found industrial applications in speciality coatings and inks. Archer Daniels Midland (ADM), an agricultural processing company who have been commercializing the production of ethyl lactate, have recently patented isoamyl lactate as a component in an environmentally friendly solvent and household cleaner.[17]

Ethyl lactate has a boiling point of 154 °C and melting point of −26 °C. It has the potential to replace many toxic halogenated solvents. A study of its physical properties neat and mixed with water was recently performed;[18] at room temperature it has a polarity (E^N_T) of ∼0.64, refractive index 1.41 and density 1.02 g cm^{-3}.

Possibly due to the presence of both ester and alcohol functional groups, ethyl lactate has been exploited very little in synthetic chemistry. However, it has been used to prepare magnetic tapes in combination with THF, successfully replacing the methyl ethyl ketone (butan-2-one) and toluene that are normally used.[19]

γ-Valerolactone (GVL) is another bio-renewable ester with potential uses as a solvent.[20] It has a low melting point ($-31\,°C$), high boiling point ($207\,°C$), high open cup flash point ($96\,°C$) and a density of $1.05\,g\,cm^{-3}$. It is miscible with water and biodegradable. Interestingly, Horvath and co-workers were able to establish that its vapour pressure is very low even at high temperatures, only $3.5\,kPa$ at $80\,°C$. It does not form an azeotrope with water and therefore water can be removed by distillation, as can volatile organic components because of GVL's low volatility and high boiling point. Its high boiling point may also be advantageous in some reaction chemistry by allowing increased rates of reaction. It is stable in air (no peroxides could be detected after 35 days) and it did not hydrolyse in water. However, it can be hydrolysed and ring-opens in the presence of acid to give γ-hydroxy-pentanoic acid, and with aqueous sodium hydroxide it forms γ-hydroxylpentanoate.

5.2.2.1 Biodiesel

Biodiesel can be derived from a variety of plant oils or animal fats including rapeseed, soybean, and even waste vegetable oil. Other crops that show promise include mustard, flax, sunflower, canola (rape) and even algae. It consists of monoalkyl esters, mainly methyl esters (MEs), of long chain fatty acids which are obtained through transesterification of the triglycerides with an alcohol, which is usually methanol (Figure 5.7). Recently, non-fuel uses of biodiesels are becoming more widespread (Table 5.4).

In particular, methyl soyate (the biodiesel formed from soybean oil and methanol) is finding industrial applications including cleaning and degreasing technologies (Table 5.5). In industry, solvents are needed to dissolve a material for its removal or transport and then are often evaporated to restore the original material. Therefore, two important parameters are *solvent power* and evaporation rate. One way to measure solvent power is the kauri–butanol value (KBV), which is a measure of the solubility of kauri gum in the solvent. A high

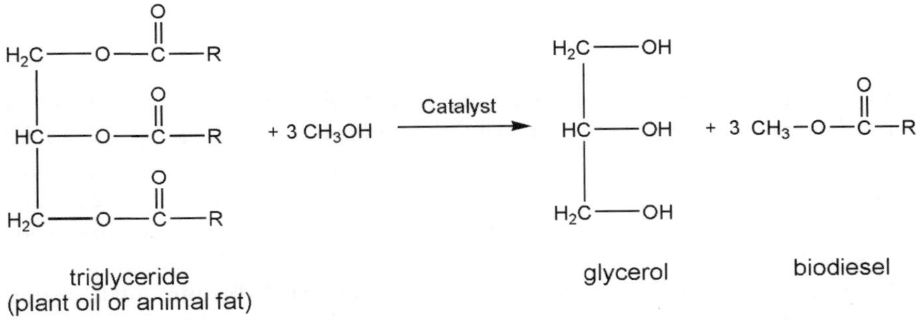

Figure 5.7 Synthesis of biodiesel.

Table 5.4   Summary of advantageous properties of biodiesels including methyl soyate as solvents.[22]

Safety advantages	Lower toxicity than toluene and methylene chloride, LD_{50} $17.4\,g\,kg^{-1}$
	Low vapour pressure, <0.1 mmHg
	High flash point, $>182\,°C$
Reaction and process advantages	Excellent compatibility with other organic solvents, metals and most plastics
	Low cost, 0.60 US$ L^{-1}
Environmental advantages	Can be bio-sourced from a range of feedstocks
	Readily biodegradable
	Low volatile organic compound level, $<50\,g\,mL^{-1}$
	Non-ozone-depleting compound
	Non-SARA reportable[a]

[a]Superfund Amendments and Reauthorization Act, http://www.epa.gov/superfund/index.htm

Table 5.5   Market applications of methyl soyate as a solvent.[22]

Industrial parts cleaning and degreasing	Household cleaners, food processing equipment cleaning, asphalt handling
	With ethyl lactate, as a cleaner in the aerospace and electronics industries
Resin cleaning and removal	Commercial and military paint strippers (replacing methylene chloride)
	Printing ink cleaners/Ink press washers (replacing toluene)
	Adhesive removers (replacing acetone)
	Graffiti removers (replacing mineral spirits, a mixture of hydrocarbons)
Cleaning up oil spills	Shoreline cleaner[a]
	Refinery or tank farm spills
	Cleaning reactors and storage tanks
Other	Carrier solvent in paints, stains and anti-corrosion coverings
	Consumer products including hand cleaners

[a]Listed on the EPA's national contingency plan, http://www.epa.gov/OEM/content/lawsregs/ncpover.htm

KBV indicates a high solvent/dissolving power. Methyl soyate has a KBV of 58, indicating that it is a strong solvent.[22] However, it is rarely used neat because it evaporates slowly and leaves a residual film on surfaces. Co-solvents with which it is formulated include ethyl lactate. This leads to a synergic effect between these two bio-sourced solvents. Ethyl lactate has a relatively high VOC level and low flash point. In a blend with biodiesel, these risks are reduced and the drying rate of the biodiesel increased. Economically, blending also makes sense as ethyl lactate is more expensive than biodiesel and therefore, its use as a renewable solvent is increased due to a more affordable market price.

A study was undertaken to assess the differences in solvent power depending on the oil and alcohol used to prepare the biodiesel.[23] The compositions of the biodiesels were measured using gas chromatography, which shows that 96–100% of each biodiesel is an alcohol ester with trace amounts of mono-, di- and triglycerides present. The presence of the glycerides has a detrimental effect on the biodiesel solvent power. The fatty acid profiles showed that linoleic acid ($C_{18:2}$) dominated all except the refined biodiesel derived from canola oil, which contained mainly oleic acid ($C_{18:1}$), the second largest component in the other biodiesels. However, it was found that the number and position of the double bond in the esters have little effect on the solvent power but unsaturated fatty acid esters have larger KBVs than saturated fatty esters. The length of the carbon chain of the fatty acid has a significant effect on the solvent power of the biodiesel: the longer the chain, the weaker the solvent power. The alcohol used to prepare the biodiesel also affects its solvent power: the smaller the alcohol, the higher the KBV of the biodiesel.

An extensive study on the use of soybean oil biodiesel as a renewable alternative to organic solvents has been published.[24] Partition coefficients between the biodiesel and water were determined for several organic species. These values were correlated with log P (1-octanol-water partition coefficient) values, which are widely used by analytical chemists and others in method development. It was found that solute distribution behaviour is similar to that of conventional solvent–water systems but is most similar to vegetable oils. When the partition coefficients for ionizable species were assessed, neutral species showed the highest distribution to the organic phase. Highly charged species and those which form hydrogen bonds with water tended to remain in the aqueous phase. Metal ions, including the actinide species UO^{2+}, showed significant partitioning into the biodiesel phase in the presence of extractants.

Recently, biodiesel has been used as a solvent in free radical-initiated polymerization reactions (Figure 5.8).[25] It should be noted that in contrast to polymerization reactions in some other green solvents, including scCO$_2$, there is no need to modify the initiator for reactions in biodiesel. All the resulting polymers except poly(methyl methacrylate) were soluble in the biodiesel. Lower molecular weights were obtained compared with conventional polymerization

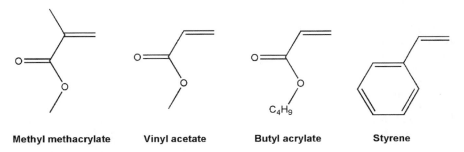

Methyl methacrylate   Vinyl acetate   Butyl acrylate   Styrene

Figure 5.8   Monomers polymerized in biodiesel.

solvents, indicating a larger degree of chain transfer in biodiesel. This may be due to C–H cleavage within the biodiesel molecule (ester) and the resulting radical being stabilized by an adjacent carbonyl group. As biodiesel has a high boiling point, polymerizations at higher temperatures can increase productivity. Interestingly, in no experiments to date has polymerization of the solvent been reported even though the biodiesel contains some C=C double bonds.

5.2.3   2-Methyltetrahydrofuran (2-MeTHF)

2-MeTHF can be made through a two-step hydrogenation of 2-furaldehyde, which can be produced using agricultural waste such as corncobs and bagasse (a by-product of the cane sugar industry).[26] The physical properties of 2-MeTHF are shown in Table 5.6, alongside some other solvents for comparison. As a substituted THF molecule, 2-MeTHF has similar properties to conventional THF, which is used in many organometallic reactions. However, as THF is miscible with water this complicates the quenching process in many of these reactions and other organic solvents have to be introduced to aid in the separation of organic and aqueous phases. In contrast, 2-MeTHF provides clean organic–water phase separations and therefore has the potential to reduce waste streams through streamlining some separation processes. It forms an azeotrope rich with water and can be more easily dried than THF or dichloromethane. It is stable to bases and in degradation studies it has been shown to be more stable towards acids than THF. In common with THF and diethyl ether, 2-MeTHF is a Lewis base and its polarity (dielectric constant and Hildebrand solubility parameter) is intermediate between these two conventional solvents. It has a higher boiling point than THF (Table 5.6), and therefore

Table 5.6   Comparison of solvent properties of 2-MeTHF with other VOC solvents.[26,27]

Property	2-MeTHF	CPME	THF	Et$_2$O	CH$_2$Cl$_2$
Dielectric constant	6.97	4.76	7.58	4.42	8.93
Density/g cm^{-3}	0.85	0.86	0.89	0.71	1.32
Boiling point/°C	80	106	65	35	40
Melting point/°C	−136	<−140	−108.5	−116	−95
Viscosity/cP	0.46	0.55	0.55	0.24	0.42
Solubility of water in solvent/g 100 g^{-1}	4.4	0.3	miscible	1.2	0.2
Azeotropic temperature with water/°C	71	83	64	34	39
Flash point/°C	−11.1	−1	−14.2	−45	na
Explosion range, lower/upper limit/vol%	1.5/8.9	1.1/9.9	1.8/11.8	1.8/48	14/22
Hildebrand/(Mpa)$^{1/2}$	16.9	–	18.7	15.5	20.2
Solvation energy/kcal mol^{-1}	0.6	–	0	2.3	–
Donor number	18	–	20.5	19.2	–

higher reaction temperatures can be used, which reduces overall reaction times. It has a low heat of vaporization, which means less solvent is lost during reaction reflux and this saves energy during distillation and recovery. Unfortunately, like most ethereal solvents, 2-MeTHF will form peroxides when exposed to air if no stabilizer is present.

Cyclopentyl methyl ether (CPME) is another alternative to typical ethereal solvents such as diethyl ether, THF, DME and dioxane.[27] At present it is not bio-sourced but it is mentioned here as it has many advantageous properties as a direct replacement for ethers. Most importantly, the rate of peroxide formation is very slow and therefore, CPME is green in terms of risk avoidance and other criteria. Its use in a range of classical and modern synthetic procedures has been reported.[27]

2-MeTHF has been used as an alternative for THF in many organometallic reactions including Grignard, Reformatskii, lithiations, hydride reductions and metal catalysed couplings. 2-MeTHF has been reported to work like THF in nickel catalysed couplings. However, in some copper-mediated couplings, 2-MeTHF gave superior diastereoselectivities compared with other solvents including THF. The highest diastereoselectivity was observed when 1,3-dinitrobenzene was used as the oxidant at −40 °C in 2-MeTHF (Figure 5.9). This procedure gave an efficient conversion to the desired biaryl in an excellent isolated yield, with no oligomers being detected.[28]

2-MeTHF has also been used as an alternative to dichloromethane in biphasic reactions including alkylations, amidations and nucleophilic substitutions.[29] For example, 2-nitrophenyl phenyl ether was prepared in 95% yield using 2-MeTHF as the organic solvent through reaction of phenol and *o*-fluoronitrobenzene using tetrabutylammonium bromide as a phase transfer catalyst.

The ability of 2-MeTHF to act as a slot-in replacement has led to its uptake in pharmaceutical process development labs.[30,31] Researchers at Eli Lilly have performed a Horner–Wadsworth–Emmons reaction using commercially available (*S*)-propylene oxide and triethylphosphonoacetate (Figure 5.10). The yield was found to be strongly influenced by the solvent used, and 2-MeTHF was found to be the superior solvent.

5.2.4 Terpenes and Plant Oils

Terpenes are a class of unsaturated hydrocarbons made up of isoprene C5 units and found in essential oils and oleoresins of plants such as conifers. The two most commonly used as solvents are turpentine and D-limonene. Their physical properties are compared with those of toluene and methylene chloride in Table 5.7. They are both immiscible with water. As can be seen in Figure 5.3, D-limonene and other small terpenes have similar molecular weights and structures to substituted cyclohexanes and toluene and are therefore to likely have solvent properties intermediate between these two VOCs.

Turpentine is a liquid obtained from the distillation of tree resin. It consists mainly of the monoterpenes α-pinene and β-pinene. As a solvent, it is used to

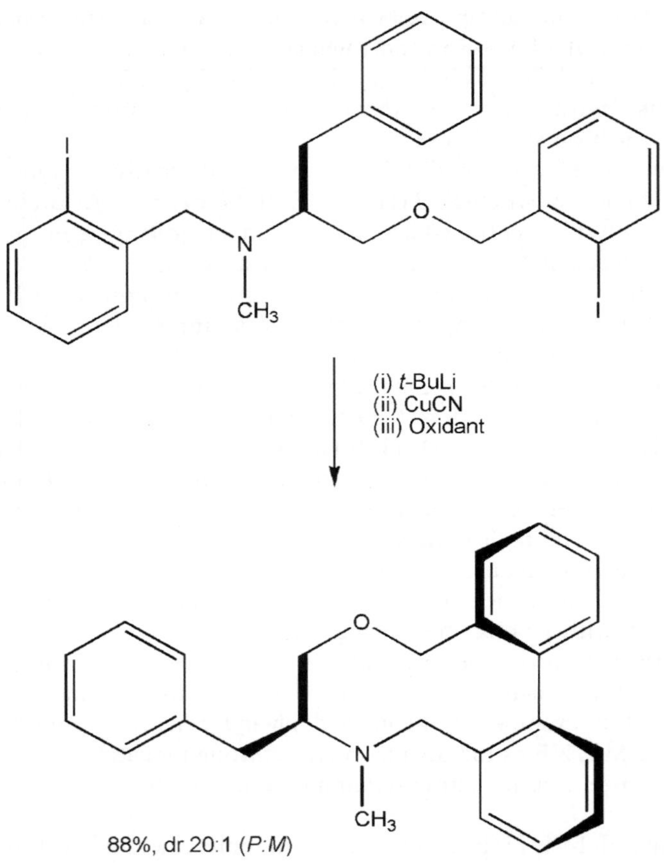

(i) *t*-BuLi
(ii) CuCN
(iii) Oxidant

88%, dr 20:1 (*P*:*M*)

Figure 5.9 Copper mediated synthesis of medium sized bi-aryl containing rings in 2-MeTHF.

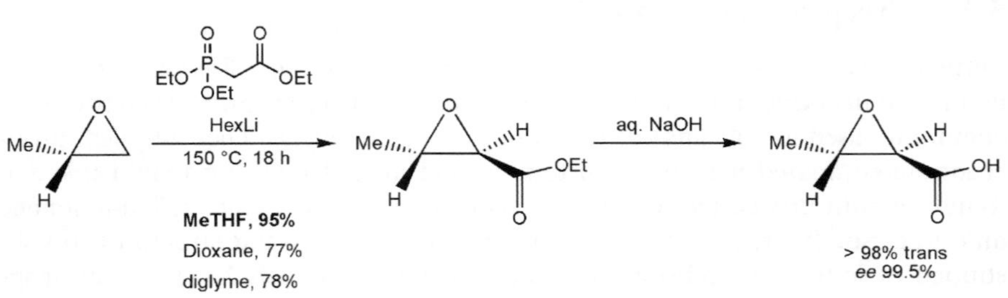

HexLi
150 °C, 18 h

MeTHF, 95%
Dioxane, 77%
diglyme, 78%

aq. NaOH

> 98% trans
ee 99.5%

Figure 5.10 Synthesis of (*R*,*R*)-2-methylcyclopropanecarboxylic acid with enhanced yields using 2-MeTHF.

Table 5.7 Some physical properties of D-limonene and turpentine alongside methylene chloride and toluene for comparison.

Property	D-Limonene	Turpentine	Toluene	CH₂Cl₂
Dielectric constant	2.37	2.2–2.7	2.38	8.93
Density/g cm^{-3}	0.84	0.85–0.87	0.86	1.32
Boiling point/°C	178	150–180	110	40
Melting point/°C	−74	< −50	−95	−95
Viscosity/cP	0.9	1.49	0.59	0.42
Vapour pressure/kPa at 20 °C	0.19	0.25–0.67	3.8	72
Flash point/°C	48	35	7	na
Explosion range, lower/upper limit/vol%	Not available	0.8/6	1.1/7.1	14/22

thin oil based paints and for producing varnishes. However, during the 20th century it was largely replaced by petroleum-sourced substitutes. Because it has a stronger and less pleasant smell than limonene, it has yet to find renewed interest as a solvent.

D-Limonene is the main component of oil extracted from citrus fruit rinds and is therefore a by-product of the fruit juice industry. Limonene can be distilled from this oil for both technical and food based uses. The extraction and distillation process is performed in all citrus-growing regions of the world to meet increasing demands for D-limonene. In particular, D-limonene is finding wide use in the manufacture of household and personal cleaning products, partly because of its pleasant aroma. It is also finding uses as an oil-rig cleaning agent, in paints, fragrance additives, cooling fluids, and other specialty products. The fact that it has poor water miscibility means that it has been employed as a floating degreaser for use in wastewater pumping stations and as a degreaser in parts washer tanks and dip tanks. In degreasing applications, its relatively high solvent power means that a smaller volume can be used. However, because of its lower volatility drying times are usually longer than for more commonly used chlorinated solvents. As a result of these diverse applications, the worldwide annual production of D-limonene is over 70 million kg and rising fast. This could lead to problems with demand outstripping supply. D-Limonene is being considered as a slot-in replacement for methyl ethyl ketone, acetone, toluene, xylene and many chlorinated solvents. However, in synthetic chemistry applications, the reactivity of the C=C double bonds must be taken into account. There are also concerns that this might lead to solvent degradation over time Table 5.8.

Limonene has recently been used in rice bran oil extraction as an alternative to hexane, which is commonly used in such processes.[32,33] The yield and quality of crude rice bran oil obtained from the limonene extraction were almost equivalent to those obtained using hexane. Interestingly, although antioxidants were not present in the limonene, only a very small amount (<1 wt%) of oxidation product was found in the recovered limonene, and therefore the solvent is potentially recyclable in such a process. This also suggests that

Table 5.8 Summary of advantageous properties of D-limonene.

Safety advantages	Lower toxicity than toluene and methylene chloride, LD_{50} 4.4 g/kg
	Relatively low vapour pressure, < 1.5 mmHg
	No known long-term health effects on humans. It is classified as non-carcinogenic and non-mutagenic. (Short-term effects include being a slight skin and eye irritant.)
Reaction and process advantages	Non-caustic and relatively inert
	Pricing competitive with conventional solvents. Food-grade limonene is twice the price of hexane.[32]
	High boiling point may be advantageous for some processes
Environmental advantages	Can be bio-sourced from a range of feedstocks
	Readily biodegradable
	Not a SARA Title III compound, and not regulated by the Clean Air Act[a]

[a]Superfund Amendments and Reauthorization Act, http://www.epa.gov/superfund/index.htm

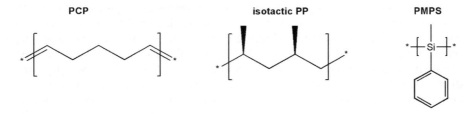

Figure 5.11   Polymers prepared in D-limonene: poly(cyclopentene) (PCP), isotactic polypropylene (PP) and poly(methylphenylsilane) (PMPS).

concerns regarding limonene degradation may currently be exaggerated. In terms of reaction chemistry, three types of polymerization reaction have been reported in D-limonene (Figure 5.11). Norbornene, 1,5-cyclooctadiene, cyclohexene and cyclopentene were polymerized by ring-opening metathesis polymerization (ROMP), using Grubb's second generation catalyst.[34] Molecular weights (M_w) of 5000–30 000 were achieved, somewhat lower than those obtained in toluene. This indicates that some side reactions occur when limonene is used as a solvent, as limonene also includes potentially reactive double bonds. When ROMP was performed in hydrogenated limonene, the M_w of the polymer increased to that observed for reactions in toluene. Therefore, side reactions when using limonene, including cross metathesis between the monomer and the vinylidene alkene of the solvent, lead to the occurrence of chain transfer and result in lower M_ws.

D-Limonene and α-pinene have been used as renewable solvents and chain transfer agents in metallocene–methylaluminoxane (MAO) catalysed polymerization of α-olefins.[35] Chain transfer from the catalyst to the solvent reduces the M_w achieved in limonene compared with toluene and also reduces the overall catalyst activity. This was confirmed, as in the ROMP studies, by performing identical reactions in hydrogenated limonene. However, an increase in stereospecificity was seen when D-limonene was used as the solvent. This is measured as the mole fraction of [mmmm] pentads seen in ^{13}C NMR spectra of the polymer. 100% isotactic polypropylene would give a value of 1.0. On performing the same propylene polymerization reactions in toluene and then in limonene, the mole fraction of [mmmm] pentads increased from 0.86 to 0.94, indicating that using a chiral solvent influences the outcome of stereospecific polymerizations. Unfortunately, when α-pinene was used, some poly(α-pinene) was found to form and this contaminates the main polymer product.

Polymethylphenylsilane (PMPS) has also been prepared via a standard Wurtz-type synthesis in D-limonene and the use of a chiral solvent has a significant effect on the M_w achieved.[36] Polysilanes are polymers with a continuous backbone of silicon atoms, which adopt helical formations in solution. They are characterized by low solubilities (that can lead to low M_w and yields) and perhaps most importantly, unique optical properties including long-wavelength UV absorption which intensifies as the degree of polymerization increases. This is associated with delocalization of the silicon–silicon sigma bonding and other orbitals. As a result, polysilanes are of interest to the opto-electronics industry. The M_w obtained in the polymerization when it was conducted at 90 °C was approximately twice that achieved when racemic limonene was used. It had been shown in previous studies that the balance of helical screw senses is the main determinant of the chain-growth polymerization mechanism for PMPS. It is clear that the chiral solvent is in some way favouring one helical screw sense over another in the growing polymer chain, and studies are ongoing in an attempt to understand this.

In summary, D-limonene has been exploited with interesting results in polymer chemistry; however, it remains to be seen whether its chirality can be used to induce similar effects in small molecule syntheses. Many of the benchmark reactions (*e.g.* Diels–Alder, Michael addition) used in the alternative solvent field are reactions of olefinic substrates and therefore could not be performed successfully in a terpene solvent.

5.2.5 Renewable Alkanes

Recently, extensive efforts have been made to synthesize liquid hydrocarbons from biomass feedstocks.[2,37–39] In 2004, Dumesic and co-workers reported that a clean stream of alkanes could be produced by aqueous phase reforming of sorbitol over a bifunctional catalyst. The sugar is repeatedly dehydrated using a solid acid catalyst and then hydrogenated using a precious metal catalyst such

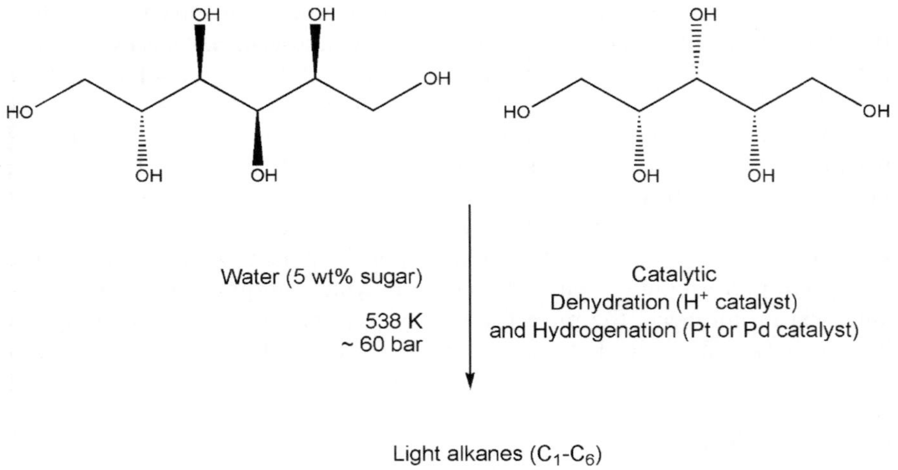

Figure 5.12   Dehydration and hydrogenation of the platform chemicals sorbitol and xylitol.

as platinum or palladium (Figure 5.12). Importantly, the hydrogen for the hydrogenation step can be made *in situ* from the sorbitol. C_1–C_6 alkanes were produced in this study and selectivity over chain length was found to vary with pH and/or the amount of solid acid added.

C_7–C_{15} alkanes can be produced through acid catalysed dehydration, followed by aldol condensation over solid base catalysts. The resulting large organic compounds are then subjected to dehydration and hydrogenation using bifunctional catalysts.[38] An aqueous feed solution is used in this process, and water is key to its success. As the organic reactant becomes hydrophobic, it can be removed from the catalyst surface using an alkane stream to prevent coke formation. This process seems to be very energy efficient and therefore could be used to produce fuel in the future. In the context of this book, the hydrocarbons could also be used as solvents. However, as with most solvents discussed in this chapter, they are flammable and hazardous and not perfect green solvents.

5.2.6   Ionic Liquids and Eutectic Mixtures Prepared from Bio-Feedstocks

Bio-sourced molecules have recently made an impact in the field of ionic liquids by yielding either the cationic or anionic moiety.[40] Many of these new solvents contain chiral centres, resulting from the abundant pool of naturally enantio-pure materials. They also frequently contain functional groups and can act as task-specific ionic liquids. A few of these are shown in Figure 5.13, but they will be discussed in more detail in Chapter 6.

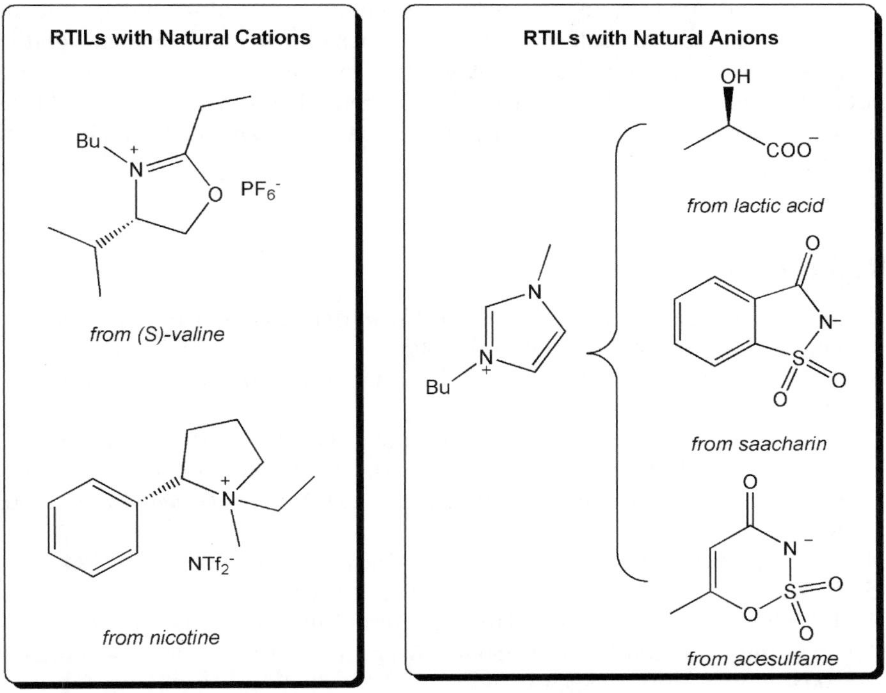

Figure 5.13 Some room temperature ionic liquids (RTILs) with naturally sourced anions or cations.

5.3 Summary and Outlook for the Future

Biotechnologists and chemical engineers have been working together for some time to develop methods for the production of a range of commodity chemicals from biomass.[41] Many of these chemicals can act as solvents, whether or not this is their intended application! Chemists are also investigating ways to catalytically deoxygenate platform chemicals and glycerol,[42] and this may lead to further bio-sourced molecules with suitable solvent properties. Additionally, many researchers are studying the catalytic conversion of cellulose directly into alcohols and alkanes. Therefore, the future looks bright for bio-sourced solvents. However, many of these solvents are still VOCs and therefore a long way from being perfect green solvents. Many are highly flammable and some are toxic. On the other hand, some are biodegradable. In the immediate future, the solvents discussed in this chapter can be used as slot-in replacements for petrochemically sourced VOC solvents. However, significant research is needed to assess the applicability of these solvents in chemical processes. For example, only in the last 2 years have reactions using glycerol as a solvent been reported. This field of greener solvents is therefore far less advanced than solvent free, water or supercritical fluids. This is in spite of the fact that in many applications a volatile medium (*e.g.* in coatings) is essential, and unfortunately not all processes that require such a solvent are amenable to more benign volatile solvents such as carbon dioxide. Also, the environmental burden of these solvents, which is already less than that of petrochemical solvents, could be further reduced in some cases if they were used in an expanded, tunable form (Chapter 9).

References

1. W. M. Nelson, *Green Solvents for Chemistry: Perspective and Practice*, Oxford University Press, Oxford, 2003.
2. J. N. Chheda, G. W. Huber and J. A. Dumesic, *Angew. Chem. Int. Ed.*, 2007, **46**, 7164.
3. J. H. Clark, V. Budarin, F. E. I. Deswarte, J. J. E. Hardy, F. M. Kerton, A. J. Hunt, R. Luque, D. J. Macquarrie, K. Milkowski, A. Rodriguez, O. Samuel, S. J. Tavener, R. J. White and A. J. Wilson, *Green Chem.*, 2006, **8**, 853.
4. G. W. Huber and A. Corma, *Angew. Chem. Int. Ed.*, 2007, **46**, 7184.
5. B. Kamm, *Angew. Chem. Int. Ed.*, 2007, **46**, 5056.
6. J. Mick, in http://www.dailytech.com/Cellulosic + Ethanol + Promises + 1 + per + Gallon + Fuel + From + Waste/article10320.htm, 14 January 2008.
7. A. Behr, J. Eilting, K. Irawadi, J. Leschinski and F. Lindner, *Green Chem.*, 2008, **10**, 13.
8. V. Budarin, R. Luque, D. J. Macquarrie and J. H. Clark, *Chem. Eur. J.*, 2007, **13**, 6914.
9. V. L. Budarin, J. H. Clark, R. Luque, D. J. Macquarrie, A. Koutinas and C. Webb, *Green Chem.*, 2007, **9**, 992.
10. A. Wolfson, C. Dlugy and Y. Shotland, *Environ. Chem. Lett.*, 2007, **5**, 67.
11. M. Pagliaro and M. Rossi, *The Future of Glycerol: New Usages for a Versatile Raw Material*, RSC Publishing, Cambridge, 2008.
12. A. Wolfson and C. Dlugy, *Chem. Pap.*, 2007, **61**, 228.
13. A. Wolfson, C. Dlugy, D. Tavor, J. Blumenfeld and Y. Shotland, *Tetrahedron: Asymmetry*, 2006, **17**, 2043.
14. M. Pagliaro, R. Ciriminna, H. Kimura, M. Rossi and C. Della Pina, *Angew. Chem. Int. Ed.*, 2007, **46**, 4434.
15. United States Environmental Protection Agency, in http://www.epa.gov/greenchemistry/pubs/pgcc/past.html, 2008.
16. Vertec-Biosolvents, http://www.vertecbiosolvents.com, accessed June 2008.
17. J. J. R. Muse, in *Environmentally friendly solvent containing isoamyl lactate*, USPTO Application #: 20070155644, U.S. Patent Office, Washington, DC, 2007.
18. S. Aparicio, S. Halajian, R. Alcalde, B. Garcia and J. M. Leal, *Chem. Phys. Lett.*, 2008, **454**, 49.

19. S. M. Nikles, M. Piao, A. M. Lane and D. E. Nikles, *Green Chem.*, 2001, **3**, 109.
20. I. T. Horvath, H. Mehdi, V. Fabos, L. Boda and L. T. Mika, *Green Chem.*, 2008, **10**, 238.
21. R. Hofer and J. Bigorra, *Green Chem.*, 2007, **9**, 203.
22. S. Wildes, *Chemical Health and Safety*, 2002, May/June, 24.
23. J. B. Hu, Z. X. Du, Z. Tang and E. Min, *Ind. Eng. Chem. Res.*, 2004, **43**, 7928.
24. S. K. Spear, S. T. Griffin, K. S. Granger, J. G. Huddleston and R. D. Rogers, *Green Chem.*, 2007, **9**, 1008.
25. S. Salehpour and M. A. Dube, *Green Chem.*, 2008, **10**, 329.
26. D. F. Aycock, *Org. Process Res. Dev.*, 2007, **11**, 156.
27. K. Watanabe, N. Yamagiwa and Y. Torisawa, *Org. Process Res. Dev.*, 2007, **11**, 251.
28. D. R. Spring, S. Krishnan and S. L. Schreiber, *J. Am. Chem. Soc.*, 2000, **122**, 5656.
29. D. H. B. Ripin and M. Vetelino, *Synlett*, 2003, 2353.
30. L. Delhaye, A. Merschaert, P. Delbeke and W. Brione, *Org. Process Res. Dev.*, 2007, **11**, 689.
31. M. Guillaume, J. Cuypers and J. Dingenen, *Org. Process Res. Dev.*, 2007, **11**, 1079.
32. P. K. Mamidipally and S. X. Liu, *Eur. J. Lipid Sci. Technol.*, 2004, **106**, 122.
33. S. X. Liu and P. K. Mamidipally, *Cereal Chem.*, 2005, **82**, 209.
34. R. T. Mathers, K. C. McMahon, K. Damodaran, C. J. Retarides and D. J. Kelley, *Macromolecules*, 2006, **39**, 8982.
35. R. T. Mathers and K. Damodaran, *J. Polym. Sci., Part A: Polym. Chem.*, 2007, **45**, 3150.
36. S. J. Holder, M. Achilleos and R. G. Jones, *J. Am. Chem. Soc.*, 2006, **128**, 12418.
37. G. W. Huber, R. D. Cortright and J. A. Dumesic, *Angew. Chem. Int. Ed.*, 2004, **43**, 1549.
38. G. W. Huber, J. N. Chheda, C. J. Barrett and J. A. Dumesic, *Science*, 2005, **308**, 1446.
39. J. O. Metzger, *Angew. Chem. Int. Ed.*, 2006, **45**, 696.
40. G. Imperato, B. Konig and C. Chiappe, *Eur. J. Org. Chem.*, 2007, 1049.
41. H. Danner and R. Braun, *Chem. Soc. Rev.*, 1999, **28**, 395.
42. M. Schlaf, *Dalton Trans.*, 2006, 4645.

CHAPTER 6

Room Temperature Ionic Liquids and Eutectic Mixtures

6.1 Introduction

Ionic liquids are defined as salts with melting points below 100 °C. They are of interest to green chemists as alternative solvents because of their inherent low volatility. However, some examples are sufficiently volatile that they can be distilled![1,2] Although the first observation of an ionic liquid occurred in 1914 ([EtNH$_3$][NO$_3$], mp 13–14 °C), it was the development of modern ionic liquids (Figure 6.1) that really accelerated research in this area during the last decade.[3]

Ionic liquids have many properties that have led to their use as reaction media and in materials processing.[4,5] They have no (or exceedingly low) vapour pressure, so volatile organic reaction products can be separated easily by distillation or under vacuum. They are thermally stable and can be used over a wide temperature range compared with conventional solvents and their properties can be readily adjusted by varying the anion and cation. For example, 1-butyl-3-methyl-imidazolium (Bmim) tetrafluoroborate (BF$_4$) is a hydrophilic solvent, whereas its hexafluorophosphate (PF$_6$) analogue is hydrophobic. The melting points of the ionic liquids are usually lower for more asymmetrical cations, e.g. [Mmim][BF$_4$], 103 °C; [Emim][BF$_4$], 6 °C and [Bmim][BF$_4$], –81 °C ([Mmim] is 1,3-dimethyl-imidazolium and [Emim] is 1-ethyl-3-methyl-imidazolium). Melting point, viscosity and conductivity data are shown in Table 6.1. In addition to their physical properties, it has also been shown that the choice of ionic liquid can dramatically affect the outcome of a chemical reaction.[6] The reaction of toluene and nitric acid was performed in three different ionic liquids. Conversions and selectivities were excellent in each case but the products were different: oxidation occurred in one case, nitration in another and halogenation in the third. In general, ionic liquids can dissolve many metal

RSC Green Chemistry Book Series
Alternative Solvents for Green Chemistry
By Francesca M. Kerton
© Francesca M. Kerton 2009
Published by the Royal Society of Chemistry, www.rsc.org

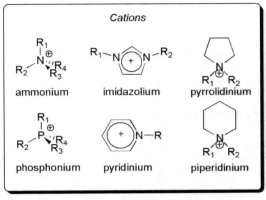

Figure 6.1 Some of the cations and anions commonly used to prepare room temperature ionic liquids (RTILs).

Table 6.1 Some physical properties of imidazolium-based ionic liquids.[4a]

Cation	Anion	Mp/°C	Thermal stability/°C	Density/ $g\ cm^{-3}$	Viscosity/ cP	Conductivity/ $ohm^{-1}\ cm^{-1}$
Emim	BF_4^-	6	412	1.24	37.7	1.4
Bmim	BF_4^-	−81	403	1.12	219	0.173
Bmim	$(CF_3SO_2)_2N^-$	−4	439	1.429	52	0.39
Bmim	PF_6^-	−61	349	1.36	450	0.146
Hmim	PF_6^-	−61	417	1.29	585	-

[a]Emim = 1-ethyl-3-methylimidazolium, Bmim = 1-butyl-3-methylimidazolium, Hmim = 1-hexyl-3-methylimidazolium

catalysts without expensive modifications, as both species are ionic, or they themselves can act as the catalytic species. Therefore, a wide range of catalytic reactions have been performed in these designer solvents, including hydrogenations, carbon–carbon bond-forming reactions and biotransformations,[7–10] and these will be discussed later. In most cases, the ionic-liquid-containing catalyst phase can be easily recycled and offers the advantages of both homogeneous and heterogeneous systems. The excellent solubility of ions in these media has also led to extensive electrochemical and metallurgical applications. In terms of greener chemistry, this is where RTILs stand out from the other alternatives, as water, carbon dioxide and renewable VOCs would just not be suitable. Additionally, RTILs are being considered as media for nuclear fuel processing; in this regard it has been shown that liquids based on 1,3-dialkylimidazolium are relatively radiation resistant.

However, there have been growing concerns over the toxicity and biodegradability of these designer solvents.[11] For example, the frequently used [Bmim][BF$_4$] and [Bmim][PF$_6$] RTILs did not show any appreciable biodegradation, whereas the exchange of the anion to octyl sulfate led to 25% biodegradation under the same conditions. Therefore, new potentially more benign ionic liquids are being developed based on non-toxic, degradable ions or at least with degradation in mind.[12–15] For example, ionic liquids containing anions derived from the sweeteners saccharin and acesulfame have properties similar to those containing the bis(trifluoromethyl)sulfonyl imide anion. In terms of toxicity[11] this anion is regarded as a risk, and it is therefore desirable that it is replaced. In terms of the toxicity of the cations, increasing the length of the additional alkyl chains on a methylimidazolium cation had a significant effect and increased toxicity.[16] For further information on the biological activity and potential risks of ionic liquids, readers are advised to read the review by Ranke and co-workers.[11]

Related to ionic liquids are substances known as *deep eutectic solvents or mixtures*. A series of these materials based on choline chloride (HOCH$_2$ CH$_2$NMe$_3$Cl) and metal chlorides, carboxylic acids or urea have been reported.[17–20] The urea–choline chloride material has many of the advantages of better-known ionic liquids (*e.g.* low volatility) but can be sourced from renewable feedstocks and is non-toxic and readily biodegradable. However, it is not an inert solvent and this has been exploited in the functionalization of the surface of cellulose fibres in cotton wool.[21] Undoubtedly, this could be extended to other cellulose based materials, biopolymers, synthetic polymers and possibly even small molecules.

The more conventional ionic liquids are generally prepared in a two-step procedure from the corresponding amine or phosphine (Figure 6.2).[22] Alkylation leads to quaternization of the heteroatom and then anion metathesis can be performed if desired. The most effective way to perform the quaternization is in solvent free conditions under microwave irradiation.[23] As they need to be prepared, RTILs are less green than many other alternative solvents in terms of life cycle

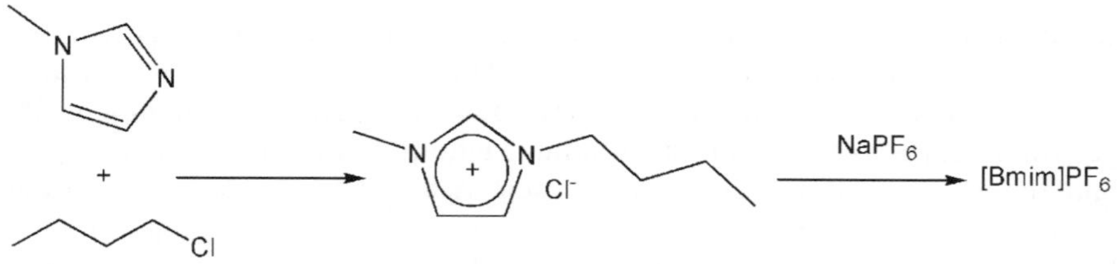

Figure 6.2   Preparation of 1-butyl-3-methylimidazolium chloride and hexafluorophosphate.

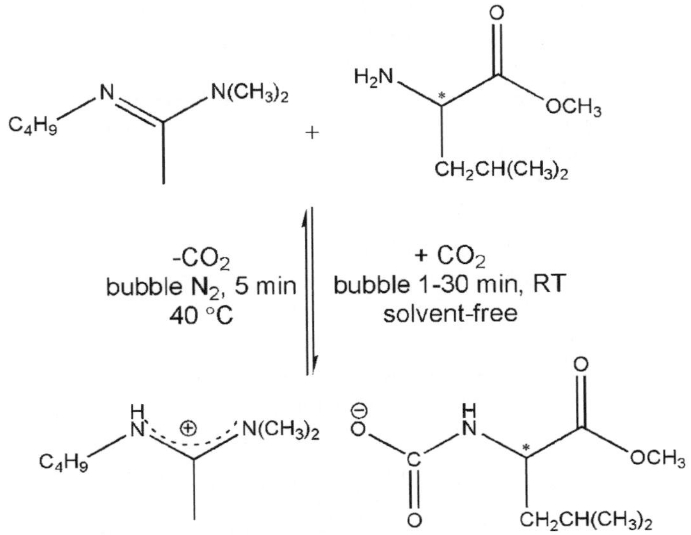

Figure 6.3 Reversible synthesis of chiral imidinium carbamate RTILs.

analysis, and this also leads to additional costs. However, they are perfect solvents for many applications. Because of their increasing popularity as solvents, several RTILs are now commercially available.[3] Further details on the synthesis of ionic liquids can be found in the books and journal articles referenced in this chapter.

Task-specific ionic liquids are becoming increasingly common; these include metal chelators[24] and chiral ionic liquids.[25] The use of chiral RTILs in synthesis will be discussed later. These can be prepared using natural chiral feedstocks, including sugars such as methyl-D-glucopyranoside.[26] However, an important new class of chiral RTIL was recently reported that was generated from the reversible reaction of amidines, amino acid esters and carbon dioxide (Figure 6.3).[27] These are a class of switchable solvents, which are further discussed in Chapter 9.

As with other solvents, it is important to consider their polarity. The polarities of ionic liquids have been measured using Reichardt's dye (Table 6.2). The commonly used [Bmim] salts have polarities close to that of ethanol. An in-depth solvatochromic study on RTIL–organic mixtures has recently been reported.[28] This work seems to indicate that despite the presence of organic co-solvents (or in synthetic chemistry, reactants) the polarity of the medium is dominated by the RTIL and the probe molecule (dye) used did not interact with the organic substance. The polarity of a RTIL can affect its miscibility (Table 6.3) and solvating power with organic compounds. In general, solids are of limited solubility in ionic liquids unless they are salts themselves, in which case they are usually very soluble. Generally, non-polar solvents such as hexane and toluene are immiscible with ionic liquids because of the extreme differences in polarity. Although dichloromethane and THF are miscible with [Bmim][PF$_6$], they may form separate phases with other RTILs. [Bmim][PF$_6$] is immiscible with water

Table 6.2 Polarities of some ionic liquids and VOCs using the E^N_T scale.[31]

Solvent	E^N_T
Hexane	0.009
[Omim][PF$_6$][a]	0.642
[Bmim][N(CF$_3$SO$_2$)$_2$]	0.642
Ethanol	0.654
[Bmim][PF$_6$]	0.667
[Bmim][CF$_3$SO$_3$]	0.667
[Bmim][BF$_4$]	0.673
Methanol	0.762
[EtNH$_3$][NO$_3$]	0.954
Water	1.000

[a]Omim = 1-octyl-3-methylimidazolium

Table 6.3 Miscibility of water and VOCs with [Bmim][PF$_6$].[4]

Solvent	ε_r	Miscibility
Water	78.3	Immiscible
CH$_3$OH	32.7	Miscible
CH$_3$CN	35.9	Miscible
Acetone	20.6	Miscible
CH$_2$Cl$_2$	8.9	Miscible
THF	7.8	Miscible
Toluene	2.4	Immiscible
Hexane	1.9	Immiscible

but, as with other RTILs, it is highly hygroscopic and therefore should be dried carefully before use. Water-miscible RTILs are more common than the immiscible ones. However, hydrophilic RTIL–water phase separation can in some cases be induced by adding a water-structuring salt such as K$_3$PO$_4$.[29] The development of more hydrophobic RTILs with anions other than the relatively unstable PF$_6$ and N(CF$_3$SO$_2$)$_2$ anions could have a significant impact on organic synthesis in RTILS. As hydrophobic RTILs allow homogeneous reactions, easy extraction of organic products and facile washing of the RTIL phase with water to remove salts and other by-products can be achieved. As discussed in Chapter 4, extensive chemistry has been performed using scCO$_2$ to extract products from RTIL–catalyst phases. Interestingly, carbon dioxide is generally very soluble in RTILs whereas hydrogen, carbon monoxide and oxygen are not. It has recently been shown that addition of carbon dioxide can enhance the solubility of gases (oxygen and methane) in some RTILs.[30]

It should be noted that RTILs do not always act as inert reaction media, and in fact many RTILs have been developed with reactivity in mind. For example, moisture-sensitive chloroaluminate ionic liquids can be used as Lewis acids and solvents simultaneously.

6.2 Chemical Examples

6.2.1 Extractions using RTILS

A short review has recently been published concerning the use of ionic liquids as extraction media. They have been used to extract or separate a wide range of substances including metal ions, organic molecules, biomolecules and gases.[32]

As a result of new rules in the EU and USA concerning transportation fuels, *e.g.* Directive 2003/17/EC, EPA 420-R-00-026, there have been significant advances in separation technologies within the oil refining industry and RTIL based technologies have emerged as leaders in this field.[33] For high levels of sulfur removal, nitrogen compounds must first be removed from diesel as they inhibit the desulfurization process. [Bmim]Cl and other chloride based RTILs have high hydrogen bond basicity. Therefore, it was possible to extract compounds containing a hydrogen-donor group such as neutral nitrogen compounds (Figure 6.4), which are difficult to remove using acetic acid treatment that can remove basic nitro-compounds. In fact, the neutral nitrogen compounds could be selectively extracted in the presence of heterocyclic sulfur compounds.[33] The ionic liquid extractant could be regenerated by treatment with water and toluene.

In the field of renewable fuels and additives, a eutectic based ionic liquid has been used to extract glycerol from biodiesel.[34] Deep eutectic solvents (DES) can be prepared from quaternary ammonium salts and a small amount of a hydrogen bond donor molecule. Glycerol is a hydrogen bond donor. Therefore, a DES for the extraction process was prepared from different quaternary ammonium salts and glycerol. The DESs formed from $EtNH_3Cl$ and $ClEt$-Me_3NCl (2-chloroethyltrimethylammonium chloride) were most effective and were able to completely remove the glycerol from the biodiesel. Initial studies towards separating the salts and glycerol by using an anti-solvent (1-butanol)

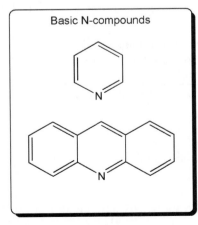

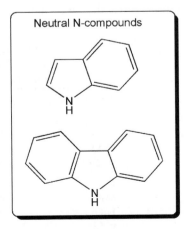

Figure 6.4 Heterocyclic aromatic nitrogen compounds found in unpurified diesel.

were promising, but work is ongoing in this area. However, such processes look like promising methods for separations in the new biodiesel industry.

1-Ethyl-3-imidazolium ethylsulfate, [Emim][EtSO$_4$], has been used as an extraction solvent for liquid–liquid extraction and as an azeotrope breaker for extractive distillation in the purification of ethyl-*t*-butyl ether (ETBE).[35] ETBE is replacing MTBE (methyl-*t*-butyl ether) as an octane booster in gasoline as it is less water soluble and therefore less likely to cause contamination through leaks. It can be prepared from ethanol, a renewable feedstock, but ETBE is very difficult to separate from ethanol because of their similar boiling points. As the cost and toxicity of [Emim][EtSO$_4$] are fairly low, and the RTIL can be recycled after use, the authors feel that this methodology could be readily applied industrially.

RTILs are also finding application in high-value pharmaceutical and bio-medical separations.[36] The traditional method used to obtain penicillin involves an organic–aqueous biphasic separation involving sequential acidification and basification. This method has several problems in addition to the use of a VOC. Proteins in the fermentation broth can cause emulsification of the separation mixture and the acidic pH used can cause penicillin decomposition. It has been reported that with the aid of a buffer salt (*e.g.* NaH$_2$PO$_4$), hydrophilic [Bmim][BF$_4$] can form an ionic liquid–aqueous two phase system (ILATPS) and can be used to selectively extract the penicillin into the ionic liquid-rich phase.[36] Upon addition of hydrophobic [Bmim][PF$_6$] to this phase, the mixture separates into two phases, a water phase containing the penicillin and a hydrophobic ionic liquid phase. The results from this study seem very promising and offer several advantages over other methods. However, it would be interesting to see if some of the more benign (more degradable, less toxic), next-generation RTILs could achieve the same goals.

Processing of metal ores is a very energy-intensive process, and the use of RTILs in this area has therefore attracted a lot of attention. Metals have been selectively extracted from mixed metal oxides using choline chloride–urea DES (Table 6.4).[37] The dissolved metals can be reclaimed using electrodeposition.

Table 6.4 Solubility of various metal oxides in a 2:1 urea–choline chloride eutectic at 60 °C.[37]

Metal oxide	Mp of metal oxide/°C	Solubility/ppm
Al$_2$O$_3$	2045	< 1
CaO	2580	6
CuO	1326	470
Cu$_2$O	1235	8725
Fe$_2$O$_3$	1565	49
Fe$_3$O$_4$	1538	40
MnO$_2$	535	493
NiO	1990	325
PbO$_2$	888	9157
ZnO	1975	8466

Other ionic liquids have also demonstrated the ability to solubilize and extract metal oxides, including protonated betaine bis(trifluoromethylsulfonyl)imide, [Hbet][N(CF$_3$SO$_2$)$_2$].[38] Soluble oxides included rare earths, uranium(VI), zinc(II), cadmium(II), mercury(II), nickel(II), copper(II), palladium(II), lead(II), manganese(II) and silver(I). Other oxides including iron(III), manganese(IV) and cobalt were insoluble or poorly soluble. Importantly, aluminium oxide and silicon dioxide were insoluble. Instead of electrochemical deposition, the metals in this study were stripped from the ionic liquid phase using an acidic aqueous solution, and the ionic liquid could be reused. Ionic liquids are also being extensively investigated as extraction media for spent nuclear fuel reprocessing, but considerable work is still required in this area.[39]

6.2.2 Electrochemistry in RTILS

Electrochemistry in RTILs has recently been reviewed,[40] and a book has been published on the topic.[41] a large number of metals have been deposited from ionic liquids (Table 6.5) and a book has also been published on electrodeposition from these media.[42] Alloys, semiconductors and conducting polymers have also been deposited from ionic liquids. The key advantages of ionic liquids for electrodeposition and electrochemical applications are their wide potential window, the high solubility of metal salts, the avoidance of water and their high conductivity compared to non-aqueous solvents.[43] There are numerous parameters that can be varied to alter the deposition characteristics including temperature, the cation and anion used, diluents and additional electrolytes.[43]

In addition to electrodeposition, ionic liquids and DESs can be used in electropolishing, which aims to remove the roughness from metallic surfaces to increase optical reflectivity for high-tech applications. For example, a eutectic mixture of ethylene glycol and choline chloride has been used in the electropolishing of various stainless steel alloys.[44] This method is preferable to current industrial procedures that use a corrosive mixture of phosphoric and sulfuric acids.

Table 6.5 Some examples of metals deposited from ionic liquids.[43]

Ionic liquid type		Metals deposited
Discrete anions	BF$_4^-$	Cd, Cu, In, Sn, Pb, Au, Ag
	PF$_6^-$	Ag, Ge
	(CF$_3$SO$_2$)$_2$N$^-$	Li, Mg, Ti, Al, Si, Ta, La, Sm, Cu, Co, Eu, Ag, Cs, Ga
Type I eutectics	AlCl$_3$	Al, Fe, Co, Ni, Cu, Zn, Ga, Pd, Au, Ag, Cd, In, Sn, Sb, Cr, Na, Li, La, Pb
	ZnCl$_2$	Fe, Mn, Ni, Cu, Co, Ti, Cr, Nb, Nd, La, Zn, Sn, Cd
Type II eutectics	CrCl$_3 \cdot$ 6H$_2$O^{20}	Cr
Type III eutectics	Urea	Zn, Sn, Cu, Ag
	Ethylene glycol	Zn, Sn

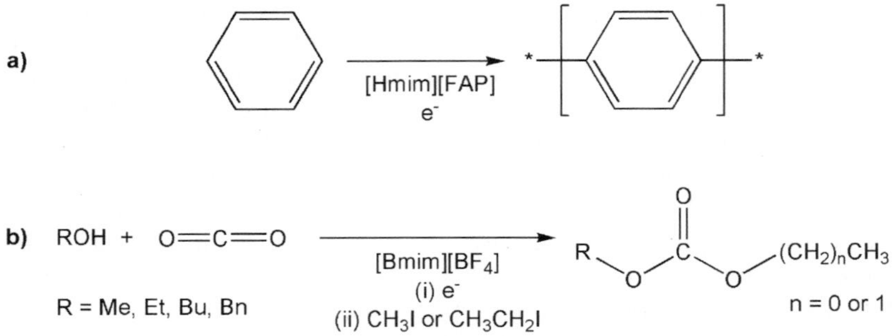

a)

b) ROH + O=C=O

R = Me, Et, Bu, Bn

[Bmim][BF₄]
(i) e⁻
(ii) CH₃I or CH₃CH₂I

n = 0 or 1

Figure 6.5 Electrochemical syntheses in ionic liquids: (a) poly(paraphenylene), (b) activation of carbon dioxide and the formation of organic carbonates.

Electrochemistry can also be used for synthesis in ionic liquids, and this is a significant advantage over many of the other solvent alternatives. Electrochemical synthetic approaches are of growing importance in green chemistry as electricity can be supplied directly from renewable resources (*e.g.* solar and wind), rather than converting electricity into heat, which reduces the overall energy efficiency. Additionally, some reactions can be performed electrochemically that cannot be performed thermally. For example, the electropolymerization of benzene has been performed in the ionic liquid 1-hexyl-3-methylimidazolium tris(pentafluoroethyl)tri-fluorophosphate, [Hmim][FAP], to yield the conducting polymer poly(paraphenylene) (Figure 6.5).[45] The resulting conjugation lengths of the polymer were between 19 and 21 and the film had a band gap of $2.9 \pm 0.2\,eV$.

Electrochemical activation of carbon dioxide has been performed in [Bmim][BF₄] (Figure 6.5).[46] This is probably very efficient because of the excellent solubility of carbon dioxide in this RTIL, and this led to very mild conditions for the activation—only 1 atm pressure was needed! This contrasts significantly with many other carbon dioxide fixations that have been reported. Additionally, no catalyst was required and the RTIL was recyclable. The scope for electrochemical reduction and fixation of carbon dioxide in RTILS is exciting and more results in this area are expected soon.

In addition to synthetic applications and the dissolution/deposition of materials, RTILs are also playing a key role in the development of new electrochemical devices including solar and fuel cells.[47]

6.2.3 Synthesis in RTILS

The range of synthetic procedures that have been performed in ionic liquids is enormous and a two-volume book has recently been published on them.[5] Therefore, only the tip of the iceberg can possibly be covered in this book on alternative solvents. In addition to their high heat capacity and the ability to perform novel separations, RTILs have allowed chemists to perform reactions that would be impossible in scCO₂ or water. For example, Grignard reagents

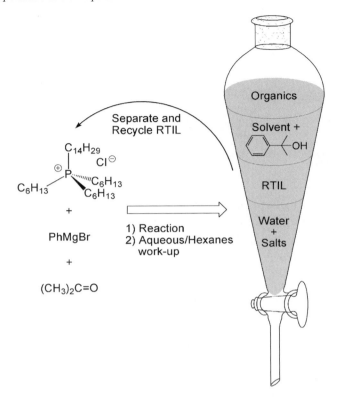

Figure 6.6 Use of Grignard reagents in phosphonium ionic liquids.

are known for their moisture-sensitive nature. However, they are versatile reagents and widely used in syntheses. Clyburne and co-workers have shown that phosphonium ionic liquids are compatible with strong bases.[48,49] For example, solutions of phenylmagnesium bromide in THF were shown to be persistent in the RTIL tetradecyl(trihexyl)phosphonium chloride for extended periods of time and could be used for many of the traditional Grignard-type reactions, including nucleophilic attack at carbonyl groups (Figure 6.6).

Some examples of organic reactions are shown in Figure 6.7. However, inorganic chemistry can also be performed in this medium. In particular, RTILs have proved themselves very effective in the stabilization of metal nanoparticles.[50,51] Stable iron, ruthenium, chromium, molybdenum, tungsten and osmium nanoparticles have been prepared by thermal or photolytic decomposition under an inert atmosphere from $Fe_2(CO)_9$, $Ru_3(CO)_{12}$, $M(CO)_6$ (M = Cr, Mo, W) and $Os_3(CO)_{12}$, dissolved in [Bmim][BF$_4$]. The particles are generally very small and uniform in size and are prepared without any additional stabilizers or capping molecules. However, it should be noted that because of the toxic nature of carbon monoxide and the metal carbonyl precursors this is far from being a particularly green reaction, but it does show the promise that

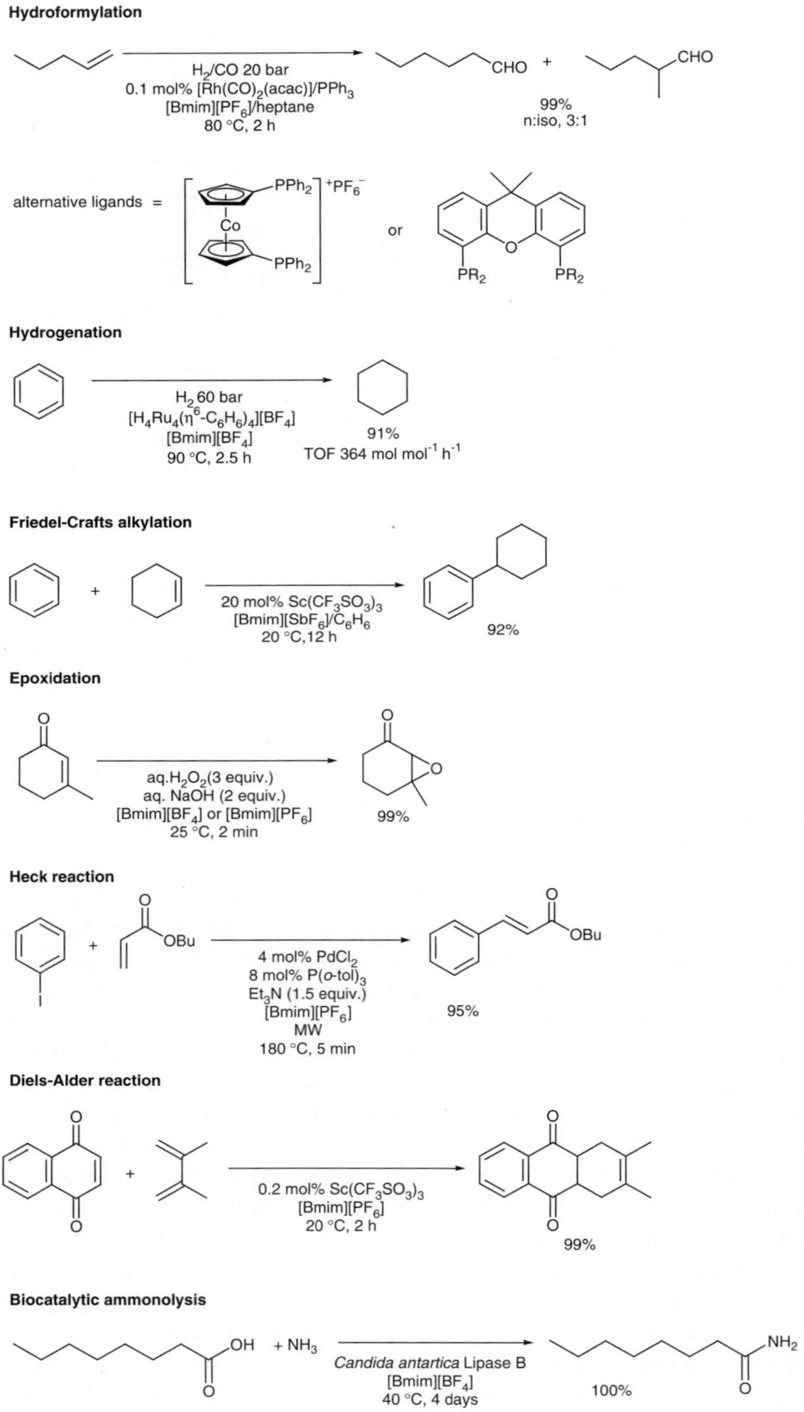

Hydroformylation
H2/CO 20 bar
0.1 mol% [Rh(CO)2(acac)]/PPh3
[Bmim][PF6]/heptane
80 °C, 2 h
CHO
+
CHO
99%
n:iso, 3:1
alternative ligands =
PPh2
+PF6−
Co
PPh2
or
O
PR2
PR2
Hydrogenation
H2 60 bar
[H4Ru4(η6-C6H6)4][BF4]
[Bmim][BF4]
90 °C, 2.5 h
91%
TOF 364 mol mol-1 h-1
Friedel-Crafts alkylation
+
20 mol% Sc(CF3SO3)3
[Bmim][SbF6]/C6H6
20 °C,12 h
92%
Epoxidation
O
aq.H2O2(3 equiv.)
aq. NaOH (2 equiv.)
[Bmim][BF4] or [Bmim][PF6]
25 °C, 2 min
O
O
99%
Heck reaction
I
+
O
OBu
4 mol% PdCl2
8 mol% P(o-tol)3
Et3N (1.5 equiv.)
[Bmim][PF6]
MW
180 °C, 5 min
O
OBu
95%
Diels-Alder reaction
O
O
+
0.2 mol% Sc(CF3SO3)3
[Bmim][PF6]
20 °C, 2 h
O
O
99%
Biocatalytic ammonolysis
OH
O
+ NH3
Candida antartica Lipase B
[Bmim][BF4]
40 °C, 4 days
100%
NH2
O

Figure 6.7 Some organic reactions in RTILs.

RTILs hold for the preparation of nanomaterials. In some cases, it is known that 'homogeneous' catalysts in RTILS act as nanocluster or nanoparticle catalysts. A recent example is an Ir(0) catalyst for acetone hydrogenation.[52]

Because of the generally excellent solubility of metal catalysts in RTILs, many of the reactions studied in these media are homogeneously metal catalysed. For example, rhodium catalysed hydroformylation reactions have been studied at length and a wide variety of phosphine ligands used. This particular reaction in RTILs has just been the subject of an extensive review.[53] In most cases, only minimal leaching of the catalyst out of the ionic liquid phase is observed and the catalysts can be very effectively recycled. These efforts are necessary because the industrial aqueous–biphasic process (Chapter 10) only works effectively for smaller olefins and therefore alternative approaches are needed for more hydrophobic, higher-mass olefins.

In addition to hydroformylation, metal catalysed hydrogenation processes have been studied at length including hydrogenation of α-olefins, aromatics and asymmetric hydrogenations of more complex substrates.[9] Benzene can be selectively fully hydrogenated by using a ruthenium cluster catalyst in [Bmim][BF$_4$].[54]

Because of the extensive amount of waste generated in traditional Friedel–Crafts reactions, it is not surprising that this reaction has been studied in RTIL. Early examples included the use of catalytic chloroaluminate ionic liquids. However, the moisture sensitivity of such systems was a drawback. Therefore, water-stable rare-earth Lewis acids, such as Sc(CF$_3$SO$_3$)$_3$, have come to be used for these reactions.[55] The same Lewis acid has also been used to catalyse Diels–Alder reactions in RTILs.[56] Interestingly, in this example, the RTIL not only provided a means for recycling the catalyst but also accelerated the rate and improved selectivity. It has also been demonstrated that a moisture stable, Lewis acidic, catalytic ionic liquid could be prepared from choline chloride and zinc dichloride, and that this was an excellent medium for the Diels–Alder reaction.[57] Yields of 90% or more were achieved in reaction times of between 8 min and 5 h for a range of dienes and dienophiles.

As with all solvent alternatives discussed in this book, palladium catalysed C–C bond-forming reactions in RTILs have been studied at length.[9,58] Because of the low volatility of ionic liquids and their rapid dielectric heating upon microwave irradiation, reaction times for Heck couplings have been significantly reduced by combining the two technologies.[59]

However, it should be noted that metals are not always used for reactions in ionic liquids and sometimes very interesting results can be obtained without them. For many epoxidation reactions, a metal catalyst, *e.g.* Jacobsen's catalyst, is used, and indeed reactions using Jacobsen's catalyst have been performed in RTILs. Interestingly, a high-yielding and quick synthesis of epoxides from electrophilic alkenes has been reported using a RTIL with aqueous hydrogen peroxide and base.[60] No hydrolysis by-products were observed and, because of the lack of additional auxiliaries, this reaction is very green.

It should be noted that many ionic liquids have some inherent basicity or acidity to their structure and can therefore act as acid or base catalysts.

For example, acetylation reactions of alcohols and carbohydrates have been performed in [Bmim]-derived ionic liquids.[61,62] If the dicyanamide anion $[N(CN)_2]^-$ is incorporated into the liquid, mild acetylations of carbohydrates can be performed at room temperature, in good yields, without any added catalyst.[62] In this example, it was shown that the RTIL was not only an effective solvent but also an active base catalyst. In a recent study, Welton and co-workers performed calculations on the gas phase basicity of the conjugate acids of possible anions from which to construct their liquid.[63] Using these data, they were able to choose the optimum RTIL in which to conduct a nucleophilic aromatic substitution reaction of an activated aniline with an activated arylhalide. Given the enormous number of possible anions and cations from which to build up an ionic liquid, the role of computation in experimental design such as this will become increasingly important.

A recent addition to the field of functional, catalytic ionic liquids comes in the area of carbon dioxide fixation. However, in this example, the reaction was performed under solvent free conditions and the ionic liquid was just used as a catalyst. Using [Bmim][OH], yields of up to 58% were obtained for the synthesis of disubstituted ureas from amines and carbon dioxide.[64] By considering the electrochemical reduction of carbon dioxide discussed above, it is clear that ionic liquids could have an important role to play in the area of carbon dioxide fixation.

In terms of chiral ionic liquids, the discussion will focus on Michael addition reactions (Figure 6.8).[25,65,66] This reaction has been performed using lactate as

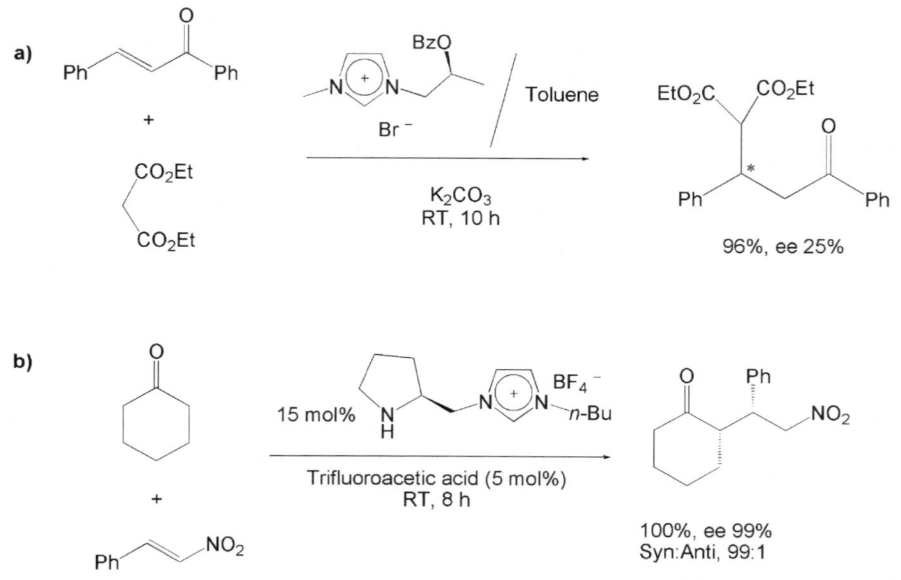

Figure 6.8 Asymmetric Michael addition reactions using chiral ionic liquids: (a) derived from lactate, (b) derived from proline.

the chiral precursor to the ionic liquid, which was obtained in around 60% overall yield.[65] The resulting chiral ionic liquid was used as the medium and chiral reagent for the enantioselective Michael addition of diethyl malonate to 1,3-diphenyl-prop-2-en-1-one (Figure 6.8a). Although the enantiomeric excess achieved was only moderate, it provided useful insights into the use of chiral ionic liquids in asymmetric induction. It should be noted that toluene was required as a co-solvent to aid in the stirring of the reaction mixture. In more recent studies a proline-derived chiral ionic liquid was prepared and employed as an efficient organocatalyst for a Michael addition of cyclohexanone to nitroalkenes. In this case, in addition to excellent conversions, excellent stereo- and enantioselectivities were achieved. However, given the multi-step proce- dure required to produce this ionic liquid, Michael additions under solvent free conditions, in water, or on water are far superior in terms of greenness. Nevertheless, chiral ionic liquids sourced from renewable feedstocks are likely to be important reaction media in a few years as the method of asymmetric induction becomes better understood.

Chiral ionic liquids have also been used to aid enantioselective metal cata- lysed reactions. For example, homogeneous rhodium catalysed hydrogenations using tropoisomeric biphenylphosphine ligands have been reported using chiral ionic liquids derived from L-proline and L-valine.[67] Enantioselectivities of up to 69% could be achieved and the catalytic system could be reused after extraction with scCO$_2$.

6.2.3.1 Biocatalysis in RTILs

Significant research efforts have been directed towards the performance of biocatalytic reactions in RTIL media and this field has recently been reviewed.[10,68] A wide range of reactions have been studied (Table 6.6), but it should be noted that most of the enzymes that have worked particularly well in RTILs are lipases.

As in their reactions in organic solvents, or for that matter scCO$_2$, the enzymes in RTILs require an optimal degree of hydration to maintain their activity. The anion component of an ionic liquid can play an important role in this regard. Therefore, anions that do not interact strongly with water are desirable for enzymatic reactions in RTILs in order to prevent water being 'stripped out' of the tertiary structure of the enzyme and solvating the anion rather than helping to maintain the activity of the enzyme. A wide range of spectroscopic techniques have been used to investigate the structures of enzymes in RTILs, including fluorescence, circular dichroism and FT-IR.[10,68] Recently, the extent of aggregation of *Candida antarctica* lipase B in a range of [Emim] ionic liquids was studied, and compared with results for the enzyme in water and DMSO, using dynamic light scattering and small angle neutron scattering techniques.[69]

A recent biocatalytic transformation has shown that just a small amount of ionic liquid may be sufficient to give some of the benefits, *e.g.* increased

Table 6.6   Some examples of biocatalysed reactions in RTILs.[68]

Enzyme class	*Reactions*	*Typical comments*
Lipase	Transesterification and direct esterification (incl. polyester synthesis) Ring-opening polymerization of ε-caprolactone Hydrolysis; alcoholysis; acetylation	Higher stability of enzyme; greater activity; catalyst recyclable; sometimes higher enantio- and regio-selectivity compared with VOCs
Esterase	Transesterification	Higher stability of enzyme; activity and enantioselectivity similar to VOCs
Protease	Transesterification Hydrolysis (incl. stereospecific)	Higher stability of enzyme; rates comparable to buffer solutions and VOCs; enhanced enantioselectivity
Dehydrogenase	Enantioselective reduction of ketone Oxidation of codeine	Faster rate than VOC
Peroxidase	Oxidation of anisoles and thioanisoles	Activity similar to VOC; stereoselectivity similar to water
β-Galactosidase (whole cells, *e.g.* baker's yeast)	Reduction of ketones	RTIL recyclable after product distilled; RTILs (alone) do not damage cell membrane

stability, that have been described for other systems (Table 6.6). By coating biocatalyst particles (*Candida antarctica* lipase B (Novozyme)) with alkyl imidazolium based ionic liquids, the activity of the catalyst towards transesterification was doubled and it was suggested that this was due to improved mass-transfer. Several citronellyl esters (acetate, propionate, butyrate, caprate and laurate) were prepared in high yields ($>99\%$) and 100% purity using equimolar mixtures of citronellol and alkyl vinyl ester as substrates under solvent free conditions.[70] The resulting terpene esters are among the most important flavour and fragrance compounds used in the food, beverage, cosmetic and pharmaceutical industries. However, although solvent use was minimized and the amount of RTIL used was small, this reaction is far from being ideally green. Vinyl acetate is carcinogenic and highly flammable, so direct esterification using an organic acid is therefore a safer and more atom-economic route to such compounds.

6.2.3.2   Polymer Synthesis and Processing

The use of RTILs in polymer synthesis and processing has significantly increased during the last decade and this has resulted in reviews and symposia dedicated to the topic.[5,71-73] However, the field is less advanced than polymerizations in other green media such as water and scCO$_2$. Reactions that have been studied in RTILs include free radical polymerizations, including styrene and alkyl methacrylates; cationic polymerization of styrene; cationic ring-opening polymerization of oxazolines;[74] reverse atom-transfer radical polymerizations, including methyl methacrylate (MMA) and acrylonitrile; ruthenium catalysed ring-opening metathesis polymerization of norbornenes;[75] and palladium catalysed copolymerization of propene with carbon monoxide.[76] Some advantages that RTILs offer for these reactions are that they are non-coordinating solvents in metal catalysed polymerizations, and in radical-initiated processes the ratio of propagation rate to termination rate is significantly higher than in conventional solvents, which can lead to the formation of very high molecular weight polymers. Some of the initial studies in this area focused on the free radical polymerization of *n*-butyl methacrylate and the effect of varying the composition of the ionic liquid upon the polymerization process.[77] Higher molecular weights were achieved than even in bulk (solvent free) polymerization processes and therefore the resulting polymers had high glass transition temperatures. The optimum ionic liquids were found to be imidazolium based rather than pyridinium and aliphatic ammonium salts. It was suggested that the high molar masses of polymers were favoured by high viscosities of the imidazolium salts and perhaps were due to locally ordered structures. It should also be noted that the ionic liquids could be recycled after the polymerization by simple decantation and extraction procedures.

More recently, the free radical polymerization of MMA in ionic liquids has been studied in depth in an attempt to understand the mechanism for the enhanced polymerization rates, high molecular weight polymers, and high yields in free radical polymerization of MMA and other methacrylate monomers.[78] Addition of a chain transfer agent in an attempt to cap (reduce) the molecular weights achieved was less effective than in conventional solvents (*e.g.* xylene) and an increase in reaction temperature to reduce molecular weight was also less effective in the RTIL [Emim][EtSO$_4$]. The researchers suggested that the radical is protected by preferentially partitioning in the ionic liquid, whereas the monomer is spread throughout the mixture in extremely small, monomer-rich domains. Excitingly, due to the protection of the radical on the growing polymer chain, the synthesis of poly(styrene)-PMMA block copolymers was possible.[78]

In addition to synthesis, RTILs are finding uses in processing of synthetic polymers, *e.g.* composite polymer–nanotube materials, as plasticizers, as porogens and in depolymerization (cracking).[73] In the field of polymer processing, if the correct two components are chosen, upon combination the polymers and ionic liquids can form ion gels (ion-conducting polymer electrolytes) and these new materials have promising electrochemical applications.[79] An excellent example of the use of reactive ionic liquids in polymer processing is in the cracking of polyethylene (PE). 1-Ethyl-3-methylimidazolium chloroaluminate has been used to break down PEs at 120 °C in the presence of a small amount of acid co catalyst such as concentrated sulfuric acid (2 mol%). This is a significantly lower temperature than is normally required (300–1000 °C). Additionally, the reaction was quite selective as the major

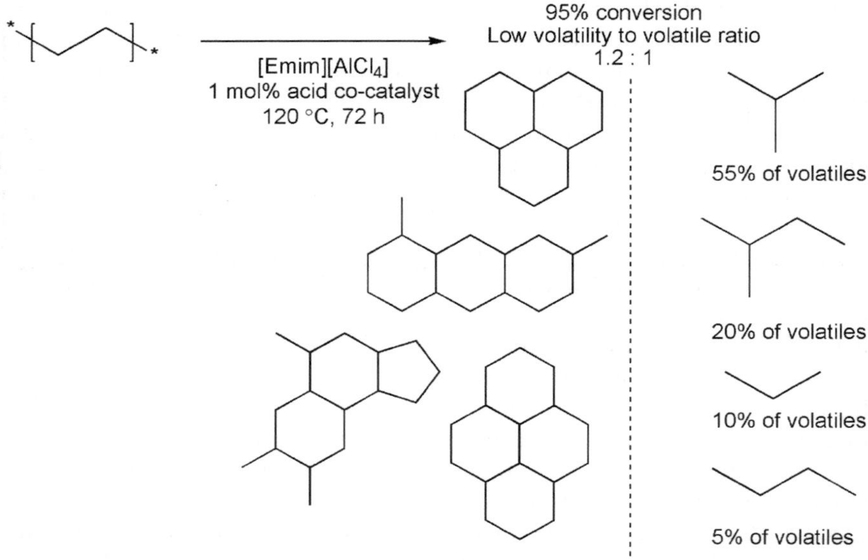

Figure 6.9   Cracking of polyethylene in a chloroaluminate ionic liquid.

products of the reaction were C_3–C_5 gaseous alkanes (such as isobutane), branched cyclic alkanes and importantly, negligible amounts of aromatics (Figure 6.9).[80] It should be noted that the reaction progresses more smoothly if finely powdered PE is used and the reaction chemistry is kept below the melting point of the polymer. This is thought to be a result of a surface area effect, whereby the molten polymer has a lower surface area in contact with the reaction medium.

However, the most exciting results have come in the area of processing natural polymers such as cellulose, lignocellulose and chitin, which are abundant and renewable materials. In 2002, Rogers and co-workers reported that cellulose from virtually any source (fibrous, amorphous, pulp, cotton, bacterial, filter paper, *etc.*) could be dissolved readily and rapidly, without derivatization, in [Bmim]Cl by gentle heating (especially with microwaves).[81] Subsequently, it was shown that the dissolved polymer could be precipitated from water in controlled architectures (fibres, membranes, beads, flocs, *etc.*) by a range of techniques. Blended and composite materials could also be formed by incorporating functional additives.[82] The additives could be soluble in the ionic liquid, *e.g.* dyes, or dispersed/insoluble, *e.g.* nanoparticles. Notably, the ionic liquid could be recycled by at least two energy-saving methods. More recently, it has been shown that other biopolymers can also be dissolved in [Bmim]Cl and that ionic liquid solutions of chitin and chitosan can reversibly adsorb carbon dioxide.[83]

Based on an understanding of how [Bmim]Cl was able to dissolve cellulose, which is generally insoluble in most common organic media, researchers have

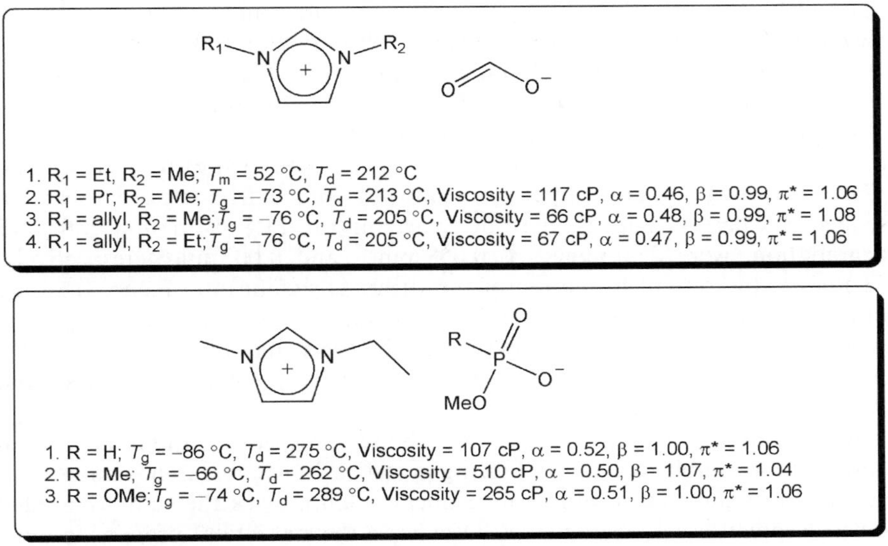

1. R_1 = Et, R_2 = Me; T_m = 52 °C, T_d = 212 °C
2. R_1 = Pr, R_2 = Me; T_g = −73 °C, T_d = 213 °C, Viscosity = 117 cP, α = 0.46, β = 0.99, π^* = 1.06
3. R_1 = allyl, R_2 = Me; T_g = −76 °C, T_d = 205 °C, Viscosity = 66 cP, α = 0.48, β = 0.99, π^* = 1.08
4. R_1 = allyl, R_2 = Et; T_g = −76 °C, T_d = 205 °C, Viscosity = 67 cP, α = 0.47, β = 0.99, π^* = 1.06

1. R = H; T_g = −86 °C, T_d = 275 °C, Viscosity = 107 cP, α = 0.52, β = 1.00, π^* = 1.06
2. R = Me; T_g = −66 °C, T_d = 262 °C, Viscosity = 510 cP, α = 0.50, β = 1.07, π^* = 1.04
3. R = OMe; T_g = −74 °C, T_d = 289 °C, Viscosity = 265 cP, α = 0.51, β = 1.00, π^* = 1.06

Figure 6.10 Structures and physical data for next-generation ionic liquids for carbohydrate dissolution.

discovered alternative, chloride-free, lower melting and less viscous RTILs that are probably preferable to [Bmim]Cl for this process. The interaction between the hydroxyl groups of cellulose and the ionic liquids is crucial for dissolution, and this has been demonstrated by methylating carbohydrates that then have a reduced solubility in the ionic liquid. The high hydrogen bonding ability of the chloride ion and the resulting interaction with the hydroxyl groups was the main reason that [Bmim]Cl could successfully dissolve cellulose. Therefore, Ohno and co-workers prepared a series of 1,3-dialkylimidazolium formates as alternative RTILs having strong hydrogen bond acceptability (Figure 6.10).[84] These formates had significantly lower viscosity than previously reported polar ionic liquids and because of their strong hydrogen bonding ability, various polysaccharides including amylose and cellulose could be dissolved in high concentrations under mild conditions. For example, 1-allyl-3-methylimidazo-lium formate could dissolve dextrin, amylose, and inulin very effectively. It should also be noted that at 3–20 wt% concentrations these solutions showed no phase separation upon cooling to room temperature and remained homogeneous even after storing at low temperatures for several months. However, upon addition of methanol or ethanol, the mixture phase separated and films or beads could be prepared. More recently, the same research group has shown that dimethyl phosphate, methyl methylphosphonate and methyl phosphonate alkylimidazolium RTILs can also dissolve cellulose under mild conditions (Figure 6.10).[85]

Another advance in this area has come in the degradation of lignocellulosic materials through hydrolysis using hydrochloric acid in an RTIL to afford improved yields of total reducing sugars (TRS) under mild conditions.[86] TRS yields of 66–81% were obtained for the hydrolysis of corn stalk, rice straw, pine wood and bagasse in [Bmim]Cl in the presence of 7 wt% hydrogen chloride at 100 °C under atmospheric pressure within 60 min. Other RTILs and acids were also studied but were not as effective as the HCl–[Bmim]Cl combination. The researchers also performed kinetic modelling based on their experimental data. The results suggest that the hydrolysis follows a consecutive first-order reaction sequence, where k_1 and k_2, the rate constants for TRS formation and TRS degradation, were determined as $0.068\,min^{-1}$ and $0.007\,min^{-1}$ respectively. Therefore, the rate of formation of the sugars is significantly higher than the rate of degradation. This exciting new system may well be valuable in providing cost-efficient conversion of biomass into biofuels and bio based products. Further exciting results are expected at the interface of bio-feedstocks and RTILs in the near future.

As well as being used to dissolve carbohydrates, ionic liquids have been designed for solubilizing and stabilizing proteins.[87] Such research is important, as many proteins that have pharmaceutical potential lack the stability needed for widespread use as therapeutics. Of course, the stability of proteins in ionic liquids is also of paramount importance for the development of biocatalytic reactions in these media. A range of biocompatible ionic liquids was found to dissolve significant amounts of the model protein cytochrome *c*. The biocompatible anions studied included dicyanamide, saccharinate and dihydrogen phosphate. The cations chosen included pyrrolidinium based cations and the biochemical cation choline. Compared to buffered aqueous solutions, the thermal stability of cytochrome *c* was dramatically increased in the dihydrogen phosphate ionic liquids, as evidenced by the disappearance of the denaturing peak from their DSC traces. Increased thermal stability was also confirmed by variable temperature ATR-FTIR spectroscopy, whereby retention of the secondary structure of the protein was confirmed. It was proposed that the nature of the anion was the important factor in these effects; the dihydrogen phosphate anion provided a proton activity similar to that in neutral water as well as hydrogen bonding donor and acceptor sites.

6.2.4   Selected Unconventional Uses of RTILs

Many uses of ionic liquids do not fit under the traditional sub-headings (synthetic chemistry, materials chemistry and extractions) used in the chapters of this book, and some of these are discussed here. In these cases, the RTIL is being used not as a solvent but more as an alternative liquid phase.

Ionic liquids have been used to prepare liquid-in-glass thermometers.[88] Traditional thermometers either contain liquid mercury, which is toxic, or ethanol, which has a boiling point of only 78 °C and so has a limited temperature range. Therefore, there is a need for an alternative if a suitable substance can be found. Two ionic liquids were chosen for study (Figure 6.11).

[TEMA][MeSO$_4$] was selected for use over an ambient temperature range ($\sim$0–100 °C) as it is reasonably inexpensive, biodegradable and has a known

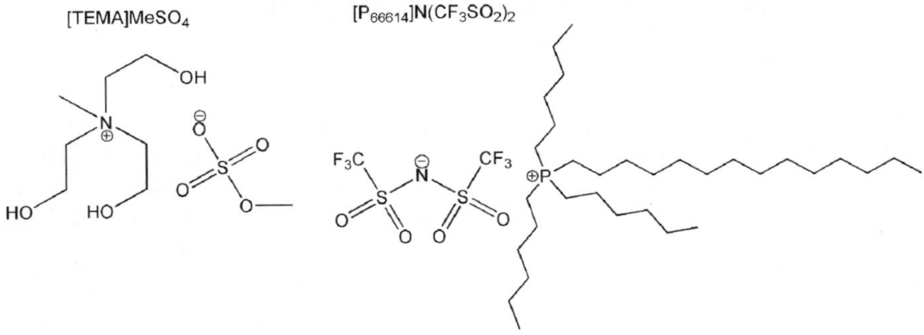

Figure 6.11 RTILs used in the preparation of new thermometers.

low toxicity. For applications requiring a wider temperature range, $[P_{66614}]$ $[N(CF_3SO_2)_2]$ was chosen because of its high decomposition temperature and low melting point. As both these RTILs are colourless, a small amount of an ionic liquid dye (1 wt%) was incorporated into the thermometers to facilitate the temperature reading. Upon calibration, these prototype thermometers worked well for measuring temperatures; however, a potential disadvantage is the relatively high viscosities and heat capacities of the RTILs, which may slow their response times. Work is ongoing to optimize these devices.

In another measurement application, ionic liquids have been used as gas sensors. These devices are becoming increasingly important for air monitoring, medical and counterterrorism applications. It has been shown that thin films of ionic liquids perform well as sensor interfaces and offer control over selectivity and sensitivity when interacting with gas phase analytes.[89] In this study, seven ionic liquids (including $[P_{66614}][MeSO_4]$) were used and provided excellent classification results for both known (100% correct) and unknown (96% correct) concentrations of organic vapours including benzene, ethanol and heptane.

One of the most unusual applications of an RTIL to date has been as the basis for a liquid mirror for a lunar telescope.[90] Liquid metallic alloys had previously been proposed for liquid mirror telescope applications; however, they were not suitable for infrared applications. 1-Ethyl-3-imidazolium ethylsulfate ([Emim][EtSO₄]) was used as the basis for the mirror. Upon coating the RTIL with silver or silver on chromium, excellent reflectivity was obtained. Unfortunately, the reflectivity of the mirror was still not sufficient for the application but these initial results are very promising and improvements to the metal film deposition are being sought, *e.g.* higher vacuum. The authors of this paper hope that the lunar liquid mirror telescope will soon become a reality and they claim it will revolutionize astronomical observations of the early universe. It is amazing to think of the huge impact that a mere 'alternative solvent' could make on an entirely different field of science!

6.3   Summary and Outlook for the Future

The field of RTILs has grown dramatically in the last 10 years and the range of anions and cations that can be used to make these non-volatile solvents is continually expanding. Therefore, calculations and mathematical modelling are required so that judicious choices can be made for this class of solvent. However, tremendous advances have been made in the field and some best guesses could be made based on these. The inertness of phosphonium based RTILs towards strong bases has opened up many avenues that could be explored further, such as catalytic coupling reactions of Grignard reagents.[91] The knowledge that RTILs containing anions with strong hydrogen bond acceptability can dissolve many carbohydrate based materials could provide an opportunity for extensive exploitation of these media in the flourishing bio-feedstock and biomaterials industries. However, it should not be forgotten that these media are more expensive than many other alternatives and carry a larger environmental burden, as they need to be synthesized. Therefore, side-by-side studies should be performed in conventional solvents, in/on water and under solvent free conditions where possible to assess whether a real advantage is obtained by using RTIL media. Nevertheless, in many cases the use of RTILs is the only viable option, such as in the growing fields of electrochemistry and electrodeposition of metals. The development of new electrolytic materials based on RTILs for fuel cells and other applications will no doubt continue to grow, especially considering the ever-increasing price of oil. Also, many metal catalysts are directly compatible with RTILs and these media allow facile recycling of these expensive chemicals, thereby significantly increasing their turnover number (TON).

During the last 5 years, tremendous efforts have been made to obtain data on the toxicity of these solvents and their persistence in the environment. However, more work is still needed in this area and collaborations with toxicologists and other scientists will be necessary to make the required advances. However, the use of natural feedstocks, *e.g.* choline chloride, in the preparation of RTILs should aid in reducing toxicity and also increase the biodegradability of these solvents. As a result of advances in this area, the label of 'green solvent' is becoming more fitting for RTILs again! And as long as the media are used in areas where they are needed, rather than being used to follow a fashion, more exciting results will surely come soon in terms of greener extractions, organic chemistry, materials chemistry, and even at interfaces with unexpected fields.

References

1. M. J. Earle, J. Esperanca, M. A. Gilea, J. N. C. Lopes, L. P. N. Rebelo, J. W. Magee, K. R. Seddon and J. A. Widegren, *Nature*, 2006, **439**, 831.
2. K. J. Fraser, E. I. Izgorodina, M. Forsyth, J. L. Scott and D. R. MacFarlane, *Chem. Commun.*, 2007, 3817.
3. N. V. Plechkova and K. R. Seddon, *Chem. Soc. Rev.*, 2008, **37**, 123.

4. D. J. Adams, P. J. Dyson and S. J. Taverner, *Chemistry in Alternative Reaction Media*, John Wiley & Sons Ltd., Chichester, 2004.
5. P. Wasserscheid and T. Welton, *in Ionic Liquids in Synthesis*, VCH, Weinheim, 2007.
6. M. J. Earle, S. P. Katdare and K. R. Seddon, *Org. Lett.*, 2004, **6**, 707.
7. J. Dupont, R. F. de Souza and P. A. Z. Suarez, *Chem. Rev.*, 2002, **102**, 3667.
8. T. Welton, *Coord. Chem. Rev.*, 2004, **248**, 2459.
9. V. I. Parvulescu and C. Hardacre, *Chem. Rev.*, 2007, **107**, 2615.
10. F. vanRantwijk and R. A. Sheldon, *Chem. Rev.*, 2007, **107**, 2757.
11. J. Ranke, S. Stolte, R. Stormann, J. Arning and B. Jastorff, *Chem. Rev.*, 2007, **107**, 2183.
12. E. B. Carter, S. L. Culver, P. A. Fox, R. D. Goode, I. Ntai, M. D. Tickell, R. K. Traylor, N. W. Hoffman and J. H. D Jr., *Chem. Commun.*, 2004, 630.
13. J. R. Harjani, R. D. Singer, M. T. Garcia and P. J. Scammells, *Green Chem.*, 2008, **10**, 436.
14. S. Bouquillon, T. Courant, D. Dean, N. Gathergood, S. Morrissey, B. Pegot, P. J. Scammells and R. D. Singer, *Aust. J. Chem.*, 2007, **60**, 843.
15. G. Imperato, B. Konig and C. Chiappe, *Eur. J. Org. Chem.*, 2007, 1049.
16. R. F. M. Frade, A. Matias, L. C. Branco, C. A. M. Afonso and C. M. M. Duarte, *Green Chem.*, 2007, **9**, 873.
17. A. P. Abbott, G. Capper, D. L. Davies, R. K. Rasheed and V. Tambyrajah, *Green Chem.*, 2002, **4**, 24.
18. A. P. Abbott, G. Capper, D. L. Davies, R. K. Rasheed and V. Tambyrajah, *Chem. Commun.*, 2003, 70.
19. A. P. Abbott, D. Boothby, G. Capper, D. L. Davies and R. K. Rasheed, *J. Am. Chem. Soc.*, 2004, **126**, 9142.
20. A. P. Abbott, G. Capper, D. L. Davies and R. K. Rasheed, *Chem. Eur. J.*, 2004, **10**, 3769.
21. A. P. Abbott, T. J. Bell, S. Handa and B. Stoddart, *Green Chem.*, 2006, **8**, 784.
22. P. A. Z. Suarez, J. E. L. Dullius, S. Einloft, R. F. DeSouza and J. Dupont, *Polyhedron*, 1996, **15**, 1217.
23. M. Deetlefs and K. R. Seddon, *Green Chem.*, 2003, **5**, 181.
24. J. R. Harjani, T. Friscic, L. R. MacGillivray and R. D. Singer, *Inorg. Chem.*, 2006, **45**, 10025.
25. A. D. Headley and B. Ni, *Aldrichimica Acta*, 2007, **40**, 107.
26. L. Poletti, C. Chiappe, L. Lay, D. Pieraccini, L. Polito and G. Russo, *Green Chem.*, 2007, **9**, 337.
27. T. Yamada, P. J. Lukac, T. Yu and R. G. Weiss, *Chem. Mater.*, 2007, **19**, 4761.
28. B. R. Mellein, S. Aki, R. L. Ladewski and J. F. Brennecke, *J. Phys. Chem. B*, 2007, **111**, 131.
29. K. E. Gutowski, G. A. Broker, H. D. Willauer, J. G. Huddleston, R. P. Swatloski, J. D. Holbrey and R. D. Rogers, *J. Am. Chem. Soc.*, 2003, **125**, 6632.

30. D. G. Hert, J. L. Anderson, S. Aki and J. F. Brennecke, *Chem. Commun.*, 2005, 2603.
31. K. Mikami, in *Green Reaction Media in Organic Synthesis*, Blackwell-Wiley, Oxford, 2005.
32. H. Zhao, S. Q. Xia and P. S. Ma, *J. Chem. Technol. Biotechnol.*, 2005, **80**, 1089.
33. L. L. Xie, A. Favre-Reguillon, X. X. Wang, X. Fu, E. Pellet-Rostaing, G. Toussaint, C. Geantet, M. Vrinat and M. Lemaire, *Green Chem.*, 2008, **10**, 524.
34. A. P. Abbott, P. M. Cullis, M. J. Gibson, R. C. Harris and E. Raven, *Green Chem.*, 2007, **9**, 868.
35. A. Arce, H. Rodriguez and A. Soto, *Green Chem.*, 2007, **9**, 247.
36. Y. Y. Jiang, H. S. Xia, C. Guo, I. Mahmood and H. Z. Liu, *Ind. Eng. Chem. Res.*, 2007, **46**, 6303.
37. A. P. Abbott, G. Capper, D. L. Davies, R. K. Rasheed and P. Shikotra, *Inorg. Chem.*, 2005, **44**, 6497.
38. P. Nockemann, B. Thijs, S. Pittois, J. Thoen, C. Glorieux, K. Van Hecke, L. Van Meervelt, B. Kirchner and K. Binnemans, *J. Phys. Chem. B*, 2006, **110**, 20978.
39. K. Binnemans, *Chem. Rev.*, 2007, **107**, 2592.
40. M. C. Buzzeo, R. G. Evans and R. G. Compton, *ChemPhysChem*, 2004, **5**, 1106.
41. H. Ohno, *in Electrochemical Aspects of Ionic Liquids*, John Wiley & Sons, New York, 2005.
42. F. Endres, D. MacFarlane and A. Abbott, *in Electrodeposition from Ionic Liquids*, VCH, Weinheim, 2008.
43. A. P. Abbott and K. J. McKenzie, *Phys. Chem. Chem. Phys.*, 2006, **8**, 4265.
44. A. P. Abbott, G. Capper, K. J. McKenzie, A. Glidle and K. S. Ryder, *Phys. Chem. Chem. Phys.*, 2006, **8**, 4214.
45. T. Carstens, S. Z. El Abedin and F. Endres, *ChemPhysChem*, 2008, **9**, 439.
46. L. Zhang, D. F. Niu, K. Zhang, G. R. Zhang, Y. W. Luo and J. X. Lu, *Green Chem.*, 2008, **10**, 202.
47. D. R. Macfarlane, M. Forsyth, P. C. Howlett, J. M. Pringle, J. Sun, G. Annat, W. Neil and E. I. Izgorodina, *Acc. Chem. Res.*, 2007, **40**, 1165.
48. T. Ramnial, S. A. Taylor, M. L. Bender, B. Gorodetsky, P. T. K. Lee, D. A. Dickie, B. M. McCollum, C. C. Pye, C. J. Walsby and J. A. C. Clyburne, *J. Org. Chem.*, 2008, **73**, 801.
49. T. Ramnial, D. D. Ino and J. A. C. Clyburne, *Chem. Commun.*, 2005, 325.
50. E. Redel, R. Thomann and C. Janiak, *Chem. Commun.*, 2008, 1789.
51. J. Kramer, E. Redel, R. Thomann and C. Janiak, *Organometallics*, 2008, **27**, 1976.
52. L. S. Ott, S. Campbell, K. R. Seddon and R. G. Finke, *Inorg. Chem.*, 2007, **46**, 10335.
53. M. Haumann and A. Riisager, *Chem. Rev.*, 2008, **108**, 1474.
54. P. J. Dyson, D. J. Ellis, D. G. Parker and T. Welton, *Chem. Commun.*, 1999, 25.

55. C. E. Song, W. H. Shim, E. J. Roh and J. H. Choi, *Chem. Commun.*, 2000, 1695.
56. C. E. Song, W. H. Shim, E. J. Roh, S. G. Lee and J. H. Choi, *Chem. Commun.*, 2001, 1122.
57. A. P. Abbott, G. Capper, D. L. Davies, R. H. Rasheed and V. Tambyrajah, *Green Chem.*, 2002, **4**, 24.
58. R. Singh, M. Sharma, R. Mamgain and D. S. Rawat, *J. Braz. Chem. Soc.*, 2008, **19**, 357.
59. K. S. A. Vallin, P. Emilsson, M. Larhed and A. Hallberg, *J. Org. Chem.*, 2002, **67**, 6243.
60. O. Bortolini, V. Conte, C. Chiappe, G. Fantin, M. Fogagnolo and S. Maietti, *Green Chem.*, 2002, **4**, 94.
61. R. Alleti, W. S. Oh, M. Perambuduru, Z. Afrasiabi, E. Sinn and V. P. Reddy, *Green Chem.*, 2005, **7**, 203.
62. S. A. Forsyth, D. R. MacFarlane, R. J. Thomson and M. von Itzstein, *Chem. Commun.*, 2002, 714.
63. I. Newington, J. M. Perez-Arlandis and T. Welton, *Org. Lett.*, 2007, **9**, 5247.
64. T. Jiang, X. M. Ma, Y. X. Zhou, S. G. Liang, J. C. Zhang and B. X. Han, *Green Chem.*, 2008, **10**, 465.
65. Z. M. Wang, Q. Wang, Y. Zhang and W. L. Bao, *Tetrahedron Lett.*, 2005, **46**, 4657.
66. S. Z. Luo, X. L. Mi, L. Zhang, S. Liu, H. Xu and J. P. Cheng, *Angew. Chem. Int. Ed.*, 2006, **45**, 3093.
67. M. Schmitkamp, D. Chen, W. Leitner, J. Klankermayer and G. Francio, *Chem. Commun.*, 2007, 4012.
68. S. Cantone, U. Hanefeld and A. Basso, *Green Chem.*, 2007, **9**, 954.
69. D. Sate, M. H. A. Janssen, G. Stephens, R. A. Sheldon, K. R. Seddon and J. R. Lu, *Green Chem.*, 2007, **9**, 859.
70. P. Lozano, R. Piamtongkam, K. Kohns, T. De Diego, M. Vaultier and J. L. Iborra, *Green Chem.*, 2007, **9**, 780.
71. P. Kubisa, *J. Polym. Sci. A*, 2005, **43**, 4675.
72. D. C. Zhao, H. T. Xu, P. Xu, F. Q. Liu and G. Gao, *Progr. Chem.*, 2005, **17**, 700.
73. C. S. Brazel and R. D. Rogers, *in Ionic Liquids in Polymer Systems: Solvents, Additives, and Novel Applications*, American Chemical Society, Washington, DC, 2005.
74. C. Guerrero-Sanchez, R. Hoogenboom and U. S. Schubert, *Chem. Commun.*, 2006, 3797.
75. Y. S. Vygodskii, A. S. Shaplov, E. I. Lozinskaya, O. A. Filippov, E. S. Shubina, R. Bandari and M. R. Buchmeiser, *Macromolecules*, 2006, **39**, 7821.
76. H. J. Wang, L. L. Wang, W. S. Lam, W. Y. Yu and A. S. C. Chan, *Tetrahedron: Asymmetry*, 2006, **17**, 7.
77. V. Strehmel, A. Laschewsky, H. Wetzel and E. Gornitz, *Macromolecules*, 2006, **39**, 923.
78. K. J. Thurecht, P. N. Gooden, S. Goel, C. Tuck, P. Licence and D. J. Irvine, *Macromolecules*, 2008, **41**, 2814.

79. T. Ueki and M. Watanabe, *Macromolecules*, 2008, **41**, 3739.
80. C. J. Adams, M. J. Earle and K. R. Seddon, *Green Chem.*, 2000, **2**, 21.
81. R. P. Swatloski, S. K. Spear, J. D. Holbrey and R. D. Rogers, *J. Am. Chem. Soc.*, 2002, **124**, 4974.
82. S. D. Zhu, Y. X. Wu, Q. M. Chen, Z. N. Yu, C. W. Wang, S. W. Jin, Y. G. Ding and G. Wu, *Green Chem.*, 2006, **8**, 325.
83. H. B. Xie, S. B. Zhang and S. H. Li, *Green Chem.*, 2006, **8**, 630.
84. Y. Fukaya, A. Sugimoto and H. Ohno, *Biomacromolecules*, 2006, **7**, 3295.
85. Y. Fukaya, K. Hayashi, M. Wada and H. Ohno, *Green Chem.*, 2008, **10**, 44.
86. C. Z. Li, Q. Wang and Z. K. Zhao, *Green Chem.*, 2008, **10**, 177.
87. K. Fujita, D. R. MacFarlane and M. Forsyth, *Chem. Commun.*, 2005, 4804.
88. H. Rodriguez, M. Williams, J. S. Wilkes and R. D. Rogers, *Green Chem.*, 2008, **10**, 501.
89. X. X. Jin, L. Yu, D. Garcia, R. X. Ren and X. Q. Zeng, *Anal. Chem.*, 2006, **78**, 6980.
90. E. F. Borra, O. Seddiki, R. Angel, D. Eisenstein, P. Hickson, K. R. Seddon and S. P. Worden, *Nature*, 2007, **447**, 979.
91. R. R. Chowdhury, A. K. Crane, C. Fowler, P. Kwong and C. M. Kozak, *Chem. Commun.*, 2008, 94.

CHAPTER 7

Fluorous Solvents and Related Systems

7.1 Introduction

7.1.1 Overview of Fluorous Approach

It has been known for some time that highly fluorinated materials (Table 7.1) are not soluble in common laboratory solvents. It is also well known that fluorinated materials such as Teflon™ are very unreactive. However, it was not until the seminal paper by Horváth in 1994[1] that the use of these materials as solvents in catalysis and separations was highlighted.[2] Since then, research in this field has flourished and there have been many specialized meetings and journal special editions dedicated to recent advances in the field.[3–6] The fluorous approach takes advantage of the low solubility of fluorinated molecules in common VOC based solvents and also their inherent lack of reactivity. Horváth coined the term *fluorous biphase system* (FBS) to describe these systems. Just as in water–organic separations, where one has an aqueous phase and an organic phase, if a highly fluorinated solvent is used, *e.g.* perfluorocyclohexane, a fluorous phase and an organic phase are seen. The appearance of these two phases is dependent on the identity of the two solvents and the temperature. This phase behaviour can be used to enable recycling of valuable catalysts and other chemicals, and allow the benefits of a heterogeneous and homogeneous system to be employed by adjusting an external variable such as temperature. Recent advances in this area will be discussed in section 7.2, following an outline of the general properties of these systems.

Various fluorous solvents are commercially available, as a result of their use in the electronics industry, and they can be obtained in a range of boiling points

RSC Green Chemistry Book Series
Alternative Solvents for Green Chemistry
By Francesca M. Kerton
© Francesca M. Kerton 2009
Published by the Royal Society of Chemistry, www.rsc.org

(Table 7.1). Perfluorinated polyethers can also be used as the fluorous phase. However, perfluorinated aromatics are usually miscible with organic solvents and therefore are not used in FBS. It should also be noted that fluorous solvents have a low solubility in water and therefore aqueous–fluorous separations can also be achieved.

In addition to their separation properties, perfluorocarbons have advantages as solvents: they are chemically unreactive, non-flammable and have a low toxicity. But their low reactivity leads to long lifetimes and as these solvents are still volatile (see boiling points in Table 7.1), there is a high chance that atmospheric contamination will occur.

As outlined in this chapter, organic solvents often do not mix with fluorous solvents and similarly organic compounds have a low affinity for these solvents and will preferentially dissolve in an organic phase. This follows the commonly used 'like dissolves like' mantra for solvents. Therefore, for catalysts and reagents to enter into a fluorous phase they are generally tagged with a fluorous label or 'ponytail', often with the general formula – $(CH_2)_n(CF_2)_{m-1}CH_3$. Many organic–fluorous solvent combinations become miscible upon heating and conversely, they separate upon cooling to give two distinct phases. Therefore, homogeneous or heterogeneous chemistry can be performed by adjusting the temperature of the system (Figure 7.1). The fluorous solvents possess densities usually between 1.7 and $1.9 \, g \, cm^{-3}$, so are more dense than common organic solvents or water. Hence, they make up the lower phase in biphasic systems.

Table 7.1 Representative fluorous solvents and physical data.[7,8]

Solvent (common name)[a]	Formula	Bp/°C	Mp/°C	Density/g cm⁻³
Perfluorooctane(s) (FC-77)	C_8F_{18}	103–105	–	1.74
Perfluorohexane (FC-72)	C_6F_{14}	57.1	−87.1	1.68
Perfluoro(methyl cyclohexane) (PFMC)	$C_6F_{11}CF_3$	75.1	−44.7	1.79
Perfluorodecaline	$C_{10}F_{18}$	142	−10	1.95
Perfluorotributylamine (FC-43)	$C_{12}F_{27}N$	178–180	−50	1.90
α,α,α-Trifluorotoluene (Oxsol-2000)	$CF_3C_6H_5$	102	−29	1.19
Perfluoropolyether (Galden HT70)	$CF_3[(OCF(CF_3)CF_2)_n (OCF_2)_m]OCF_3$ MWt 410	70	<−110	1.7–1.8
Perfluoropolyether (Galden HT110)	$CF_3[(OCF(CF_3)CF_2)_n (OCF_2)_m]OCF_3$ MWt 580	110	<−110	1.7–1.8

[a]if available

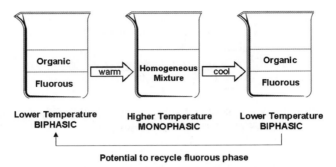

Figure 7.1 Phase separation induced by changing temperature in a fluorous–organic solvent system.

The FBS approach has now been used in many different ways, including the following.

1. *Traditional FBS* (separation by liquid–liquid extraction) (Figure 7.1).
2. *Amphiphilic solvent approach* (separation by filtration after the reaction). An amphiphilic solvent (e.g. α,α,α-trifluorotoluene, $CF_3C_6H_5$) may provide an appropriate solubility for both the fluorous and the organic materials and the reaction can proceed in a homogeneous fashion with a single solvent. After the reaction is complete, an organic solvent can be added to precipitate the fluorous material.
3. *Fluorous reverse–phase silica gel* (separation by solid phase extraction). The hydroxyl residues on silica gel are modified with perfluoroalkyl chains. This causes a fluorophilic effect between the fluorous reagent/catalyst/product and allows facile separation independent of temperature.
4. *Triphasic reactions.* For example, fluorous–organic–aqueous phases or two organic phases separated by a fluorous phase in a U-tube reaction flask.[9]
5. *Fluorous biphasic catalysis without fluorous solvents* (filtration of a thermomorphic fluorous catalyst). This can be used when a fluorous catalyst exhibits significantly different solubility in an organic solvent upon changing the temperature of the system.[10,11]

7.1.2 Fluorous Solvent Polarity Data, Solubility and Miscibility Data

Perfluorinated solvents have extremely low polarities and are generally poor solvents for commonly used organic reagents and molecules. Therefore, Reichardt's dye, which is a useful indicator for measuring the relative polarity of a solvent, is insoluble in these solvents. Specially designed fluorinated versions of this dye were not successful in obtaining the necessary data, so

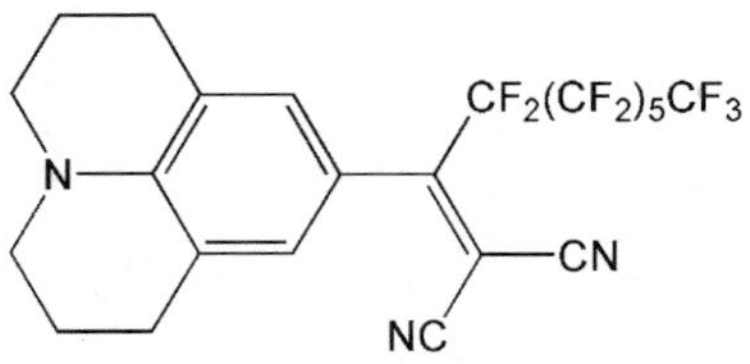

Figure 7.2 Perfluoroalkyl-substituted solvatochromic dye used to produce a spectral polarity index (P_s) for fluorous solvents.

Table 7.2 Representative perfluoromethylcyclohexane/organic solvent miscibility data.[7]

Solvent system[a]	Phase	Temperature (°C)
$CF_3C_6H_{11}$/CHCl$_3$	Two phase	RT
	One phase	$> 50.1^b$
$CF_3C_6H_{11}$/CH$_3$C$_6$H$_5$	Two phase	RT
	One phase	$> 88.6^b$
$CF_3C_6H_{11}$/hexane	Two phase	~ 0
	One phase	RT^c
$CF_3C_6H_{11}$/ether	Two phase	~ 0
	One phase	RT^c

[a]Data for perfluoromethylcyclohexane is shown as this is the generally preferred fluorous solvent for exploratory and mechanistic studies.
[b]Consulate temperature.
[c]Experimental observation, not a consulate temperature.

comparative E^N_T or $E_T(30)$ values are not available. However, a different fluorinated dye molecule (Figure 7.2) allows a spectral polarity index (P_s) to be obtained.[12,13] Perfluorocarbons are much less polar than their analogous alkanes. For example, for perfluoro(methylcyclohexane) (PFMC) $P_s = 0.46$ and for methylcyclohexane $P_s = 3.34$. On the other hand, fluorinated alcohols have higher P_s values than non-fluorinated alcohols, suggesting a stronger ability to hydrogen bond. Tables of P_s values can be found in the *Handbook of Fluorous Chemistry*.[2]

Kamlet–Taft parameters have also been obtained for some perfluorinated solvents.[14] They are not hydrogen bond donors (α is typically 0.0) and are typically extremely poor hydrogen bond acceptors (β is small and negative). They are extremely difficult to polarize, and this leads to a strongly negative π^* parameter. This last property explains why they typically form biphasic systems with organic solvents.

Although the biphasic properties of fluorous–organic systems are desirable for separations, monophasic conditions would favour enhanced reaction rates. Therefore, it is important to know the general miscibilities of fluorous solvents and the effect of temperature (Tables 7.2 and 7.3). In Table 7.2, the temperature given for the phase separation is a 'consulate' or 'upper critical solution' temperature. However, these temperatures should only be taken as a guide, as

Table 7.3 Partition coefficients for some organic and fluorous compounds in FBS.[7]

Substance	Solvent system	Partitioning % organic–fluorous
$CH_3(CH_2)_8CH_3$	$CH_3C_6H_5/CF_3C_6F_{11}$	$94.6/5.4^a$
$CH_3(CH_2)_{14}CH_3$	$CH_3C_6H_5/CF_3C_6F_{11}$	$98.9/1.1^a$
Cyclohexanol	$CH_3C_6H_5/CF_3C_6F_{11}$	$98.4/1.6^a$
C_6F_6	$CH_3C_6H_5/CF_3C_6F_{11}$	$72.0/28.0+$
$CF_3(CF_2)_7(CH_2)_3NH_2$	$CH_3C_6H_5/CF_3C_6F_{11}$	$30.0/70.0^a$
$[CF_3(CF_2)_7(CH_2)_3]_2NH$	$CH_3C_6H_5/CF_3C_6F_{11}$	$3.5/96.5^a$
$[CF_3(CF_2)_7(CH_2)_3]_3N$	$CH_3C_6H_5/CF_3C_6F_{11}$	$0.3/99.7^a$
$[CF_3(CF_2)_7(CH_2)_3]_3P$	$CH_3C_6H_5/CF_3C_6F_{11}$	$1.2/98.8^a$
$[CF_3(CF_2)_5CH_2CH_2]_3SnH$	$MeOH/CF_3C_6F_{11}$	$3.0/97.0^b$
$[\{CF_3(CF_2)_5(CH_2)_2\}_3P]_3RhCl$	$CH_3C_6H_5/CF_3C_6F_{11}$	$0.14/99.86^c$

Methods used to determine partitioning:
^{a}gas chromatography,
^{b}gravimetric,
^{c}inductively coupled plasma-atomic emission spectroscopy.

in most situations more than two components will be present, which will affect the phase behaviour and may decrease the temperature at which a single phase is observed. Also, certain solutes may cause a 'salting-out' effect and increase the temperature required to form a monophasic solution.

So why do two layers form when the mixing of two liquid phases is entropically favourable? Enthalpy must be the dominant thermodynamic driving force in the phase separation. Intermolecular attractive interactions in the pure fluorous phase (low-polarity medium) are very weak compared to the interactions in the pure non-fluorous (organic) phase. (The weak intermolecular forces in perfluorinated molecules were previously discussed in Chapter 4.) When the fluorous and organic phases mix, the stronger intermolecular interactions between the organic molecules become diluted, and the slight increase in intermolecular interactions between the fluorous molecules and the now present organic molecules is not sufficiently large to counteract the enthalpy change occurring as a result of the dilution effect. Therefore, the two liquid phases do not mix. The same approach can be used in an attempt to understand the solubility and partitioning of different compounds in FBS. Some representative partitioning values are given in Table 7.3 and a certain amount of 'leaching' into the organic or fluorous phase can be seen in all examples. Therefore, it is important to assess such effects in your own chemistry, and practical guidelines have been developed for this to aid in consistency when comparing experiments from different research groups.[7] Considerable work has recently been undertaken by Curran and co-workers to tune fluorous systems and increase partition coefficients.[15,16] Figure 7.3 can be used as a starting point in choosing phases that are likely to separate or in choosing a co-solvent that will render a miscible pair immiscible. For example, HFE-7100 and DMF are miscible, but adding FC-72 to the mixture renders the medium more fluorophilic and the DMF phase separates.[15] Conversely, adding water renders the

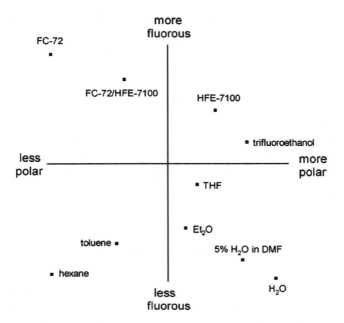

Figure 7.3 Qualitative representation of various solvent systems used in fluorous biphasic separations. [Reprinted with permission from *Org. Lett.*, 2005, 7, 3677–3680. Copyright 2005 The American Chemical Society.]

medium more fluorophobic and a HFE-7100 phase separates. The effect of water in this regard has previously been seen in fluorous solid phase extraction and HPLC applications, and therefore water, another green solvent, can often be used to encourage phase separations in this field.

Tuning to increase solubility of some non-fluorous compounds in fluorous media has recently been reported. Krytox, a commercially available poly(perfluoroether) lubricant, has a carboxylic acid end group and when this is added to an FC-72 phase, extraction of substituted pyridines is enhanced.[17] This has been attributed to the formation of a hydrogen bond complex. This study shows that non-covalent interactions can be used in modifying the fluorous phase and tailoring its properties for a particular separation.

An interesting effect is seen when FBSs are exposed to pressures of carbon dioxide. Pressures of between 16 and 50 bar can cause many such systems to become monophasic at room temperature and this may have applications in future separations.[18] Related to this is the use of carbon dioxide pressure as a switch for recycling a fluorous catalyst on a fluorinated silica support.[19]

Gases are often thought to be much more soluble in a fluorous phase than in organic solvents. This misapprehension has perhaps come about as a result of the extensive research into artificial bloods, which contain perfluorinated compounds.[20] In terms of mole ratios, oxygen is five times more soluble in perfluoromethylcyclohexane than in THF, and hydrogen 4.5 times more soluble.[7] However, given the higher molecular weight of perfluorinated solvents and higher densities compared with conventional organics, the molal concentrations are actually fairly similar. Therefore, rate enhancements involving gaseous reagents are not a foregone conclusion in fluorous media.

7.1.3 Fluorous Catalysts and Reagents

Perfluorinated molecules are prepared from their hydrocarbon analogues by electrochemical fluorination or by fluorination using cobalt trifluoride. Functional perfluorinated molecules are then used to prepare the tagged catalysts and reagents (Figure 7.4). Therefore, in terms of life cycle analysis, fluorous solvents are not as green as a solvent that does not need to be prepared, *e.g.* water, or a solvent that requires little substrate modification, *e.g.* a renewable VOC. However, the ability of FBSs to perform efficient separations often reduces the overall amount of solvent that is required in a process and therefore they are considered green alternative solvents.

Some molecules that are compatible with fluorous media are shown in Figure 7.5; they typically contain at least one $-C_6F_{13}$ or longer perfluoroalkyl chain. These fluorophilic molecules can be thought of as being designed as three component species. The fluorous group is attached to an organic group, phenyl or aliphatic $-(CH_2)_n-$, which acts to 'shield' the functional (or reactive) group from the electron-withdrawing effect of the perfluoroalkyl group. The functional group can therefore be whatever a chemist desires it to be, from a protecting group to a scavenger or a catalyst. Further examples of these fluorophilic molecules will be seen in section 7.2.

Another recent advance in this area is the development of fluorous ionic liquids.[21,22] These can contain perfluoroalkyl groups in the cation or in the

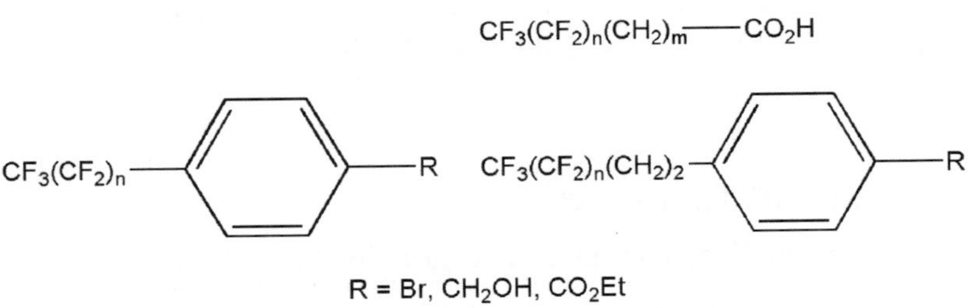

Figure 7.4 Commercially available fluorous building blocks for the preparation of fluorous tagged substrates, reagents and ligands.

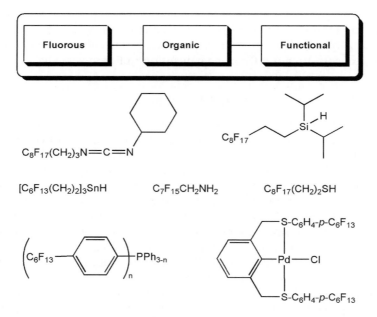

Figure 7.5 Representative examples of perfluoro-alkyl substituted molecules used in fluorous media.

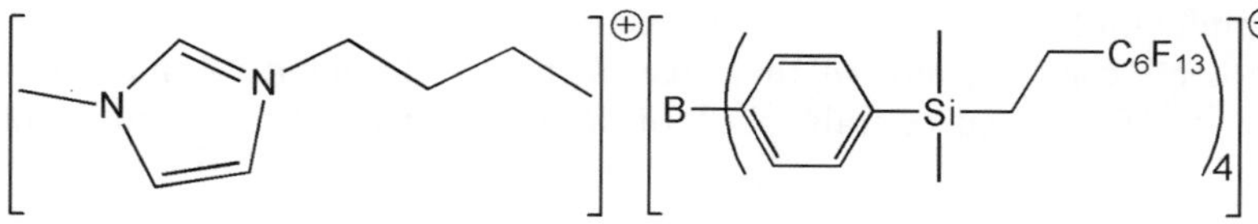

Figure 7.6 A fluorous ionic liquid containing a perfluoroalkyl-substituted borate anion.

anion. An ionic liquid containing fluorous groups within the anion is shown in Figure 7.6. Although it is an ionic liquid and possesses some of the properties and advantages of such a solvent, it also possesses phase-separation behaviour with organic solvents typical of a fluorous medium.

7.2 Chemical Examples

7.2.1 Fluorous Extractions and Fluorous Analytical Chemistry

One of the first examples of the use of fluorous solvents in reactions was their use in the extraction of photodegraded solid and liquid wastes contaminated with polychlorinated biphenyls (PCBs).[23] Fluorinated ligands and scavengers

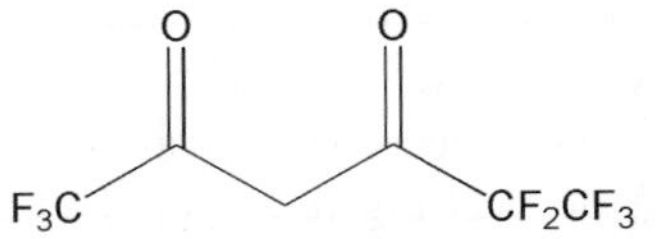

Figure 7.7 Perfluorinated β-diketone used in metal extraction studies.

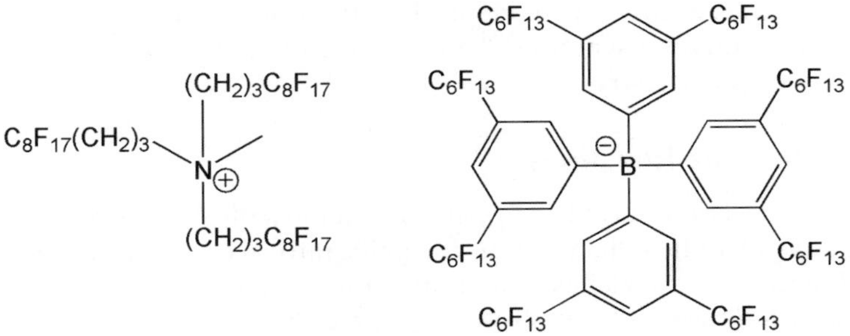

Figure 7.8 A fluorous ammonium borate electrolyte salt.

can also be used to extract trace metals from organic reaction mixtures by forming fluorous metal complexes.[23] Recently, liquid–liquid extractions using FC-72 have been performed and allowed the selective extraction of metal ions from both aqueous and organic phases.[24] Fe^{3+} could be quantitatively and selectively extracted from an acetonitrile solution containing Fe^{3+}, Co^{2+}, Ni^{2+} and Cu^{2+} using a perfluorinated β-diketone ligand (Figure 7.7). Analytical chemistry experiments have also recently shown that perfluorinated solvents with ether groups undergo a small but measurable association with monocations such as Na^+.[25] This has cast doubt on some earlier assumptions concerning fluorous solvents containing heteroatoms (oxygen and nitrogen). In the course of this study, Bühlmann and co-workers also developed a fluorophilic electrolyte salt (Figure 7.8) that may find applications in battery technology and fuel-cell research. The sodium salt of the fluorous borate shown in Figure 7.8 has since been used in the assembly of fluorous pH electrodes.[26]

Recently a method has been described in the patent literature for the fractionation of essential oils using a fluorinated solvent.[27] Oils studied included clove bud and bergamot, and fractionations could be performed in a semicontinuous mode. Fluorous solvents, as very non-polar media, offer an interesting alternative to the aqueous or alcoholic solvent approaches typically used in natural product extractions. However, it is unlikely that the technique will become widely used in this field.

A significant amount of effort in fluorous analytical chemistry is directed towards fluorous HPLC and new fluorous silicas for the separation of fluorous molecules. However, it should be noted that fluorous molecules sometimes interact sufficiently with conventional silicas that standard chromatographic techniques can be used in their separation, including traditional or reverse phase chromatography. When fluorous silica is used the separations capitalize on the ability of fluorous solid phases to separate molecules by fluorine content. Compounds lacking the fluorous tag (*e.g.* –C$_7$H$_{15}$) come off with the solvent front, as do most other non-fluorinated organic compounds. The fluorinated compounds then elute from the column strictly in order of fluorine content and a solvent gradient is sometimes needed to push the most highly fluorinated members of a series off the column. Fluorous HPLC has been successfully applied to the separation of a complex library of organic compounds prepared using fluorous mixture synthesis,[28] and is therefore a very powerful separation tool in fluorous chemistry.

7.2.2 Fluorous Reactions

Diels–Alder reactions have been performed in most alternative reaction media. For certain substrates this reaction is significantly accelerated in fluorous solvents (Figure 7.9).[29] This has been ascribed to a *fluorophobic effect*, analogous to the better-known hydrophobic effect where there is an inverse relationship between reaction rate and the solubility of reagents. However, it should be noted that in general cycloaddition reactions (including Diels–Alder reactions) are faster in water and this can be attributed to additional hydrogen bond stabilization of the transition state.

More recently, a fluorous organocatalyst has been used to perform selective Diels–Alder reactions of dienes with α,β-unsaturated aldehydes in acetonitrile–water.[30] The chiral fluorous imidazolidinone catalyst can be recovered using fluorous silica (80–90% recovery efficiency) and reused, Figure 7.10. Further organocatalytic reactions are presented later in this chapter.

In addition to fluorous media being used directly with unmodified reagents in organic synthetic procedures, more extensive use of the fluorous biphase concept has been made by using organic reagents with fluorous ponytails or fluorous reagents which can facilitate the purification of the product. This is shown schematically in Figure 7.11. They are particularly useful where by-products cause particular difficulties in reaction work up and usually lead to heavily contaminated products; for example, a fluorous Mitsunobu reaction is shown in Figure 7.12.[31] A review of fluorous approaches to organic synthesis has recently been published and would be a good starting point for chemists considering this approach in their procedures.[32]

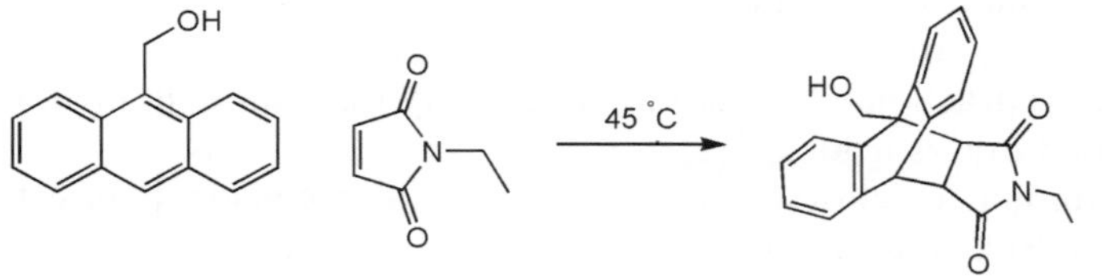

Solvent	k_{rel}
CH$_3$CN	1.0
EtOH	7.8
FC-75	42.2
C$_6$F$_{14}$	49.5

Figure 7.9 Rate enhancement of a Diels–Alder reaction in fluorous media.

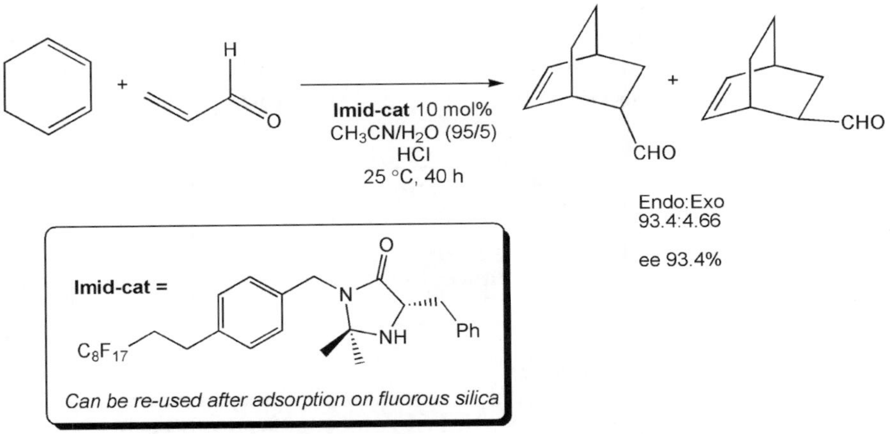

Figure 7.10 Stereospecific Diels–Alder reaction using an organocatalyst and fluorous silica for catalyst recovery.

7.2.3 Fluorous Biphase Catalysis

In 1994, Horváth and Rábai reported the first fluorous biphase catalytic system.[1] They studied hydroformylation of olefins and demonstrated the extraction of their rhodium catalyst, which contained the trialkyl phosphine $P(CH_2CH_2C_6F_{13})_3$, from the organic toluene phase. The reaction could be performed in a semi-continuous fashion with the fluorous catalyst-containing phase being reused nine times to give a total TON in excess of 35 000. Additionally, in the presence of a large amount of phosphine (PR_3:Rh, 103:1), a good ratio of linear to branched aldehyde isomers was achieved. There have been many equally elegant studies in this field since this initial report and this chapter will focus on some of the more recent advances.

Catalytic reactions that have been studied to date under FBS conditions include hydrogenations, hydroborations, hydrosilations, carbon–carbon bond formations and oxidations of sulfides, alkenes, alkanes and aldehydes. Many of these reactions can be performed in an asymmetric (or enantioselective) fashion by employing a suitable chiral ligand. Therefore, chiral fluorous ligands have been developed including a version of the widely used **BINAP** ligand (Figure 7.13). In asymmetric catalysis, the ligand is often more expensive than the precious metal and therefore there is a strong motivation to recycle such species. Horn and Bannwarth have used the fluorous BINAP shown here in Ru(II) catalysed asymmetric hydrogenation of olefins and were able to successfully reuse the catalyst through means of noncovalent immobilization on fluorous silica gel.[33] This also allowed the reaction to be performed in methanol, avoiding the use of expensive fluorous solvents. However, in some cases α,α,α-trifluorotoluene had to be added to obtain optimal stereoselectivity. Also of note are the low levels of ruthenium in the product using this method (1.6–4.5 ppm *vs* 300 ppm using conventional methods).

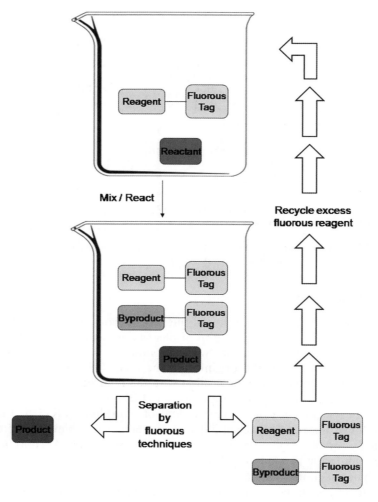

Figure 7.11 Use of fluorous reagents/tags in synthesis to aid in product isolation.

Olefin metathesis reactions are an extremely valuable class of synthetic methods. One of the most widely used catalysts for olefin metathesis is Grubbs's ruthenium carbene complex shown in Figure 7.14. Many fluorous versions of this complex have been studied as a result.[34] The air-stable fluorous complex shown can be prepared in moderate yields at room temperature in a straightforward ligand substitution reaction of a suitable precursor species with a fluorinated phosphine in trifluoromethylbenzene. Several other analogues could also be prepared with slightly differing fluorous phosphines. The complex shown in Figure 7.14 is described as being moderately fluorophilic. Although the fluorous phosphine itself has a partition coefficient of $>99.7:<0.03$ for $CF_3C_6F_{11}$–toluene at 25 °C, the ruthenium complex is in fact much more

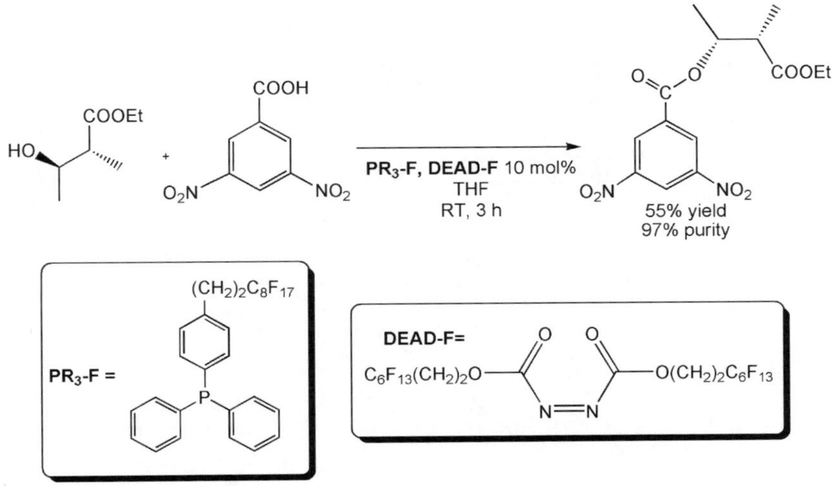

Figure 7.12 Mitsunobu reaction using fluorous tagged reagents.

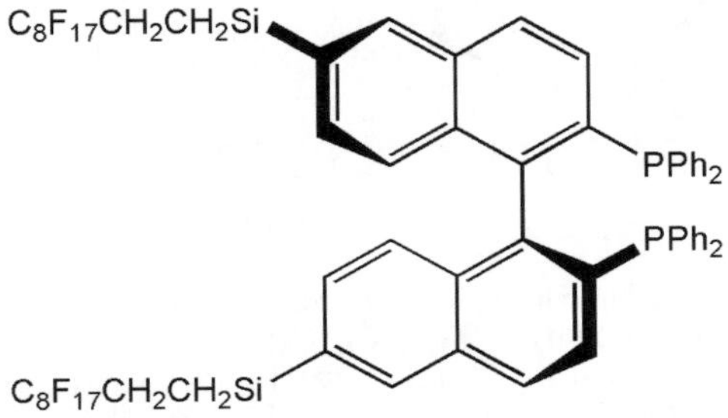

Figure 7.13 Perfluoroalkyl tagged chiral (*S*)-BINAP ligand.

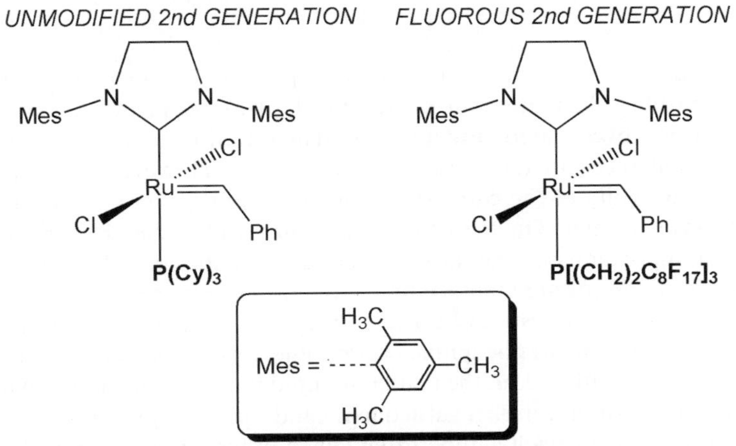

Figure 7.14 Comparison of conventional Grubbs' second generation metathesis catalyst (left) and Gladysz's fluorous version (right).

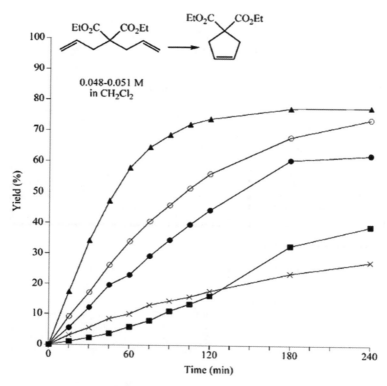

Figure 7.15   RCM in a fluorous system. Rates of formation of cyclopentene product. Solvent systems: ▲ $CH_2Cl_2/C_8F_{16}O$ (2.2 mL/1.1 mL); ○ $CH_2Cl_2/CF_3C_6F_{11}$ (4.0 mL/2.0 mL); ● $CH_2Cl_2/CF_3C_6F_{11}$ (5.0 mL/2.5 mL); × $CH_2Cl_2/CF_3C_6F_{11}$ (2.2 mL/1.1 mL); ■ CH_2Cl_2 (3.1 mL) ($C_8F_{16}O =$ perfluoro(2-butyltetrahydrofuran)). [Reprinted with permission from *Adv. Synth. Catal.*, 2007, *349*, 243–254. Copyright 2007 Wiley-VCH.]

soluble in the organic phase and exhibits a partition coefficient of 39.6:60.4. Rates of product formation in prototypical ring-closing metathesis (RCM) reactions were investigated (Figure 7.15). The initial rate of formation of the product is enhanced in the presence of $CF_3C_6F_{11}$. However, because of the significant solubility of the catalyst in organic solvents, there are difficulties in recycling such species. On the other hand, this study does demonstrate that there is an opportunity to use fluorous media to aid in the formation of active, coordinatively unsaturated metal complexes. The phosphine ligand in Grubbs's second-generation catalyst dissociates to form the catalytically active species. Therefore, if the fluorous phosphine becomes sequestered in the fluorous phase, the vacant coordinate site on the active ruthenium catalyst should remain open for substrate coordination and subsequent catalytic transformation. It remains to be seen if this approach can be extended to other types of catalyst.

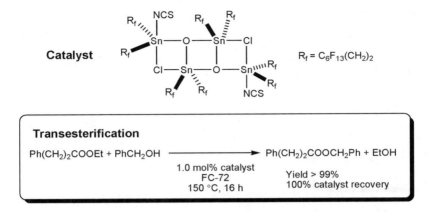

Figure 7.16 Representative transesterification using a fluorous stannoxane catalyst in a single fluorous solvent system.

Novel Lewis acidic tin catalysts have been developed for esterification reactions under fluorous conditions (Figure 7.16).[35] They can be performed in a single fluorous solvent, a binary fluorous–organic solvent system or a single organic solvent system. The catalysts employed could be recycled at least 10 times without any loss in reactivity. Also of note is that in direct esterification reactions, selective esterification of aliphatic carboxylic acids can occur in the presence of aromatic ones. The 1,3-disubstituted tetrafluoroalkyldistannoxanes are described as having a dimeric formulation and this leads to a metaloxane core that is surround by eight fluoroalkyl groups, making the surface of the catalytic molecule very fluorophilic. For condensation reactions, such as direct esterification, the use of fluorous solvents that are hydrophobic allows the reaction to be driven to completion without the need for any dehydrating agent. This technology could potentially be applied to other condensation reactions.

Fluorous rhodium complexes such as $ClRh[P(CH_2CH_2C_8F_{17})_3]_3$ are excellent catalysts or precatalysts for the hydrosilation of carbonyl compounds.[36] TONs between 100 and 500 are achieved, depending on the solvent and substrate under investigation. The catalysts can be efficiently recycled up to four times under organic–fluorous liquid–liquid biphasic conditions and TONs are maintained during each run. These catalysts can also be successfully recycled three times in the absence of fluorous solvents using Teflon tape as the delivery and recovery medium (Figure 7.17). It is proposed that attractive interactions are in operation between the fluorous domains of the catalyst and the tape. A change in temperature is used to trigger readsorption of the catalyst on to the tape when the reaction is complete.

Easy recycling of gold hydrosilation catalysts has also been achieved using a fluorous approach.[37] Conversions varied from moderate to excellent for the reaction of dimethylphenylsilane with benzaldehyde. However, the mechanism is not clear at this stage. The catalyst could not be recycled in the absence of fluorous solvents under thermomorphic conditions and the formation of

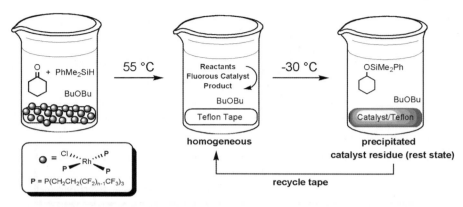

Figure 7.17   Recycling of a thermomorphic fluorous rhodium hydrosilylation catalyst
using Teflon tape. [Reprinted with permission from *Angew. Chem. Int.
Ed.*, 2005, **44**, 4095–97. Copyright 2007 Wiley-VCH.]

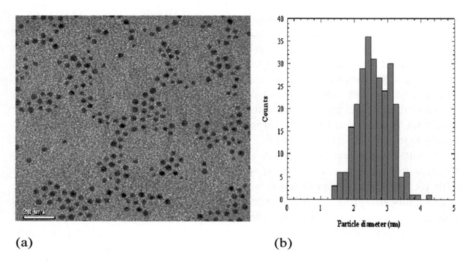

(a)   (b)

Figure 7.18   Catalytic gold nanoparticles: (a) TEM image, (b) particle size distribution of particles: distribution (%) *vs* diameter of particles (nm) [Rep-
rinted with permission from *QSAR Comb. Sci.*, 2006, **25**, 719–722.
Copyright 2006 Wiley-VCH.]

narrow polydispersity gold nanoparticles was observed (Figure 7.18). Supported fluorous phase catalysis is becoming increasingly popular with or
without fluorous solvents. Teflon-supported catalysts for hydrosilation have
been discussed above. However, in rhodium catalysed hydrogenation the
highest reaction rates were observed using fluorous mesoporous silica (up to 1.5
times faster than Teflon) and in general during the hydrogenation significantly
less rhodium leaching occurred using this support than with Teflon.[38]

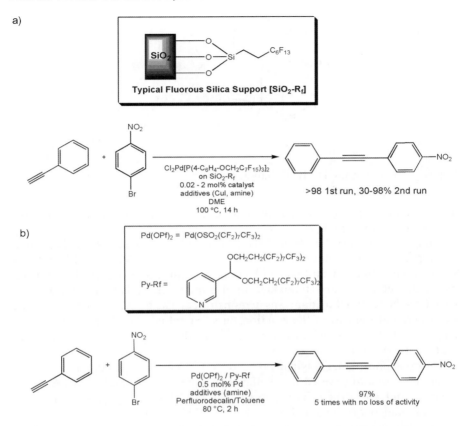

Figure 7.19 Fluorous Sonogashira reactions: (a) using a fluorous support and no fluorous solvent, (b) using perfluorodecaline under phosphine- and copper-free conditions.

Fluorous soluble or supported nanoparticles have also been used in catalytic carbon–carbon bond-forming reactions.[39] Palladium-mediated Suzuki and Sonogashira couplings have been performed using supported fluorous phase catalysis without the need for expensive perfluorinated solvents (Figure 7.19a).[40,41] An additional advantage of such a method is easy handling of small amounts of catalyst as a result of dilution with the support material. Unfortunately, in many cases when the catalyst was reused the yields were not as good for second or third runs. Interestingly, this was overcome to some extent by using water as the reaction medium rather than DME.[40] Another recyclable catalyst system for Sonogashira couplings was recently reported (Figure 7.19b).[42] A novel fluorinated palladium source, Pd(OPf)$_2$, was prepared from palladium carbonate and heptadecafluorooctanesulfonic acid and was used with a pyridine ligand bearing two fluorous ponytails. The reaction was performed under an air atmosphere using phosphine- and copper-free conditions. The catalytic systems could be repeated five times following separation

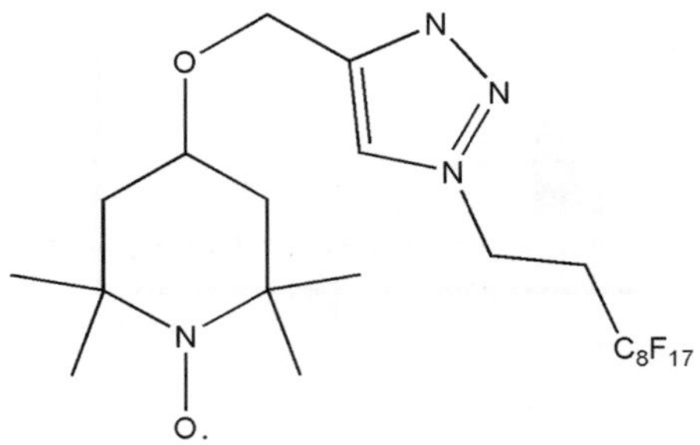

Figure 7.20   Fluorous TEMPO catalyst.

and recycling of the fluorous phase containing the palladium–perfluoro-alkylated pyridine catalyst.

A copper catalysed click (azide–alkyne cycloaddition) reaction has been used to prepare a fluorous-tagged TEMPO catalyst (Figure 7.20).[43] TEMPO is a stable organic free radical that can be used in a range of processes. In this case, its use in metal-free catalytic oxidation of primary alcohols to aldehydes using bleach as the terminal oxidant was demonstrated. The modified TEMPO can be sequestered at the end of the reaction on silica gel 60 and then released using ethyl acetate for reuse in further reactions; in this way the TEMPO was used four times with no loss in activity.

Fluorous phosphines originally developed for metal catalysed reactions have themselves been discovered to be efficient catalysts for a number of processes. Gladysz and co-workers have shown that $P[(CH_2)_2C_8F_{17}]_3$ can catalyse the addition of alcohols to methyl propiolate.[10,11] This process can also be catalysed by $P(n\text{-}Bu)_3$ in conventional solvents; however, when the fluorous phosphine is used it can be recycled using standard liquid fluorous biphase conditions or through its thermomorphic behaviour in octane. Another way to recycle the catalyst is through its adsorption on Teflon beads or shavings. Organocatalysed reactions are a growing area of catalysis chemistry and therefore more fluorous biphase organocatalytic systems will be discovered in due course.

Another recent addition to the fluorous biphase toolbox is the discovery of fluorous phase transfer catalysts for halide substitution reactions in aqueous–fluorous systems.[44] This class of reactions is academically intriguing, as an ionic displacement reaction has taken place in one of the least polar solvents known. They make use of fluorous phosphonium salts under biphasic conditions but can also make use of non-fluorous phosphonium salts in a triphasic system. Further information and reactions using such systems will no doubt be reported in the next few years.

7.2.3.1   Continuous Fluorous Biphase Catalysis

Given the moderately high cost of fluorous solvents and modified catalysts, in order to optimize the benefits of FBS the development of systems that could be

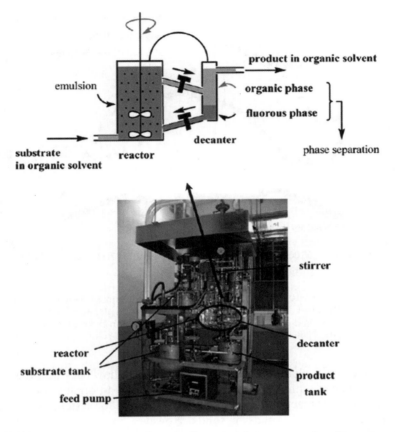

Figure 7.21 Bench scale continuous flow reaction system for fluorous–organic biphasic chemistry. [Reprinted with permission from *QSAR Comb. Sci.*, 2006, **25**, 697–702. Copyright 2006 Wiley-VCH.]

used on a continuous basis was imperative. Nishikido and co-workers had developed a range of lanthanide(III) bis(perfluoroalkanesulfonyl)amides which were highly active catalysts in Lewis-acid promoted reactions including esterifications, Diels–Alder additions, and Baeyer–Villiger and Friedel–Crafts reactions. Using an FBS approach, the catalysts were readily recyclable. Therefore, in an effort to reduce the amount of fluorous solvent required and further increase TON for their catalysts, they devised a continuous-flow system (Figure 7.21).[45,46] Conversions could be maintained at a high level for over 500 h and excellent TON achieved. This relatively simple engineering idea takes lanthanide Lewis acid catalysis and fluorous reactions out of the realm of clever academic chemistry and into the real world where the expensive components can now be efficiently reused.

A vigorously stirred reactor produces an emulsion of the two phases. This mixture flows into a decanter where the organic and fluorous phases are allowed to separate. The organic phase, which contains the product, is removed.

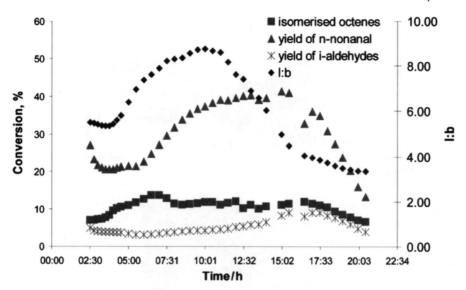

Figure 7.22   Results for the continuous hydroformylation of 1-octene catalysed by
Rh/P(4-$C_6H_4C_6F_{13}$)$_3$ in fluorocarbon solvents. (l = linear isomer, b =
branched isomer) [Reprinted with permission from *Dalton Trans.*, 2004,
2062–2064. Copyright 2004 The Royal Society of Chemistry.]

The fluorous lower phase is recycled by allowing it to flow back into the reactor
where it combines with a fresh supply of substrate in an organic solvent.

The continuous reactor described above works well for systems where all the
reagents are dissolved liquids or solids. However, many industrial processes
involve the transformation of gaseous feedstocks: one such reaction is hydroformylation. A continuous reactor has since been designed for this reaction
under FBS conditions and operated for 20 h with full catalyst recycling.[47]
A combined gaseous pressure of 15 bar was used, and typical results are shown
in Figure 7.22. A rise in conversion can be seen over the period 5–12 h. It has
been proposed that this is due to some leaching of the phosphine into the
organic phase, as some phosphine oxide was collected as a white precipitate in
the product phase and the reaction is known to be negative order in phosphine.
Unfortunately, the linear:branched (l:b) ratio was reduced in this reactor
compared to batch reactions. However, the reaction does compare favourably
with commercial rhodium catalysed propene hydroformylation that has rates in
the region 500–700 h^{-1}, whereas this process has an average rate of 750 h^{-1} and
the catalyst TON was > 15 500.

7.2.4   Fluorous Biological Chemistry and Biocatalysis

The use of fluorous solvents in catalysis has recently moved into the realm of
biocatalysis.[48] Protein–surfactant complexes were formed by hydrophobic ion

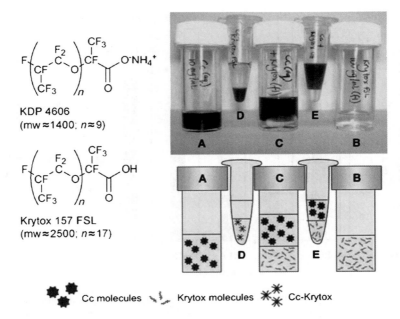

Figure 7.23 Hydrophobic ion pairing of cytochrome c (Cc) with fluorinated surfactants KDP or Krytox: (A) Dark aqueous solution of the haem protein, Cc. (B) Krytox dissolved in PFMC. (C) A biphasic system is initially observed with Cc in the aqueous (top) phase. (D) On stirring, Cc is extracted into the lower fluorous phase as it forms ion pairs with Krytox molecules. (E) If Krytox alcohol (no acidic group) is used, ion pairing is not possible and Cc stays in the aqueous phase. Note: Cc and Krytox molecules are not drawn to scale. HIP complexes with only one Cc molecule surrounded by Krytox molecules are shown for clarity. [Reprinted with permission from *Angew. Chem. Int. Ed.,* 2007, **46**, 7860–7863. Copyright 2007 Wiley-VCH.]

pairing between a highly fluorinated anionic surfactant and cytochrome c (Figure 7.23). This solubilized up to 20 mg(protein) mL^{-1} in PFMC. Interestingly, this approach could also be used to solubilize proteins in scCO$_2$. Circular dichroism spectra of the fluorous mixtures showed that the protein retained its α-helical secondary structure. Dynamic light scattering measurements show that small aggregates of protein molecules are surrounded by surfactant molecules. Following on from this study, biocatalytic reactions were performed using the enzyme α-chymotrypsin in the transesterification of *N*-acetyl-L-phenylamine with *n*-propanol and the enzyme maintained its activity over four reaction cycles.

Previously, Beckman and co-workers had prepared nicotinamide adenine dinucleotide (NAD) with a fluorophilic ponytail (FNAD).[49] This molecule was able to act as an affinity surfactant and extract the enzyme horse liver alcohol dehydrogenase (HLADH) from an aqueous medium into methoxynonafluorobutane (HFE) (Figure 7.24). Interestingly, the addition of potential

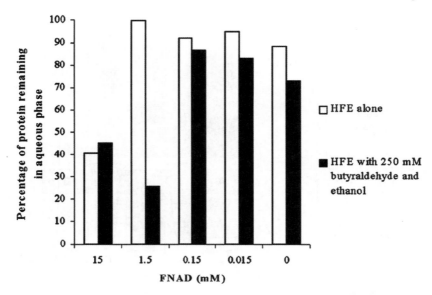

Figure 7.24   Extraction of HLADH into HFE by FNAD. [Reprinted with permission
from *Chem. Commun.*, 2002, 928–929. Copyright 2002 The Royal Society
of Chemistry.]

substrates for the enzyme (butyraldehyde and ethanol) allowed a lower concentration of the FNAD to be used and perform the phase transfer effectively.
The resulting HFE phase was not fully transparent, suggesting that an emulsion
containing large micellar species had formed. Notably, the emulsions were
stable for over 1 month. Some initial catalytic studies were also reported, *e.g.*
butyraldehyde was successfully converted to butanol by the enzyme-containing
fluorous phase.

7.2.5   Fluorous Combinatorial Chemistry

The rapid synthesis of a range of chemicals in parallel, or combinatorial
chemistry, is a growing area of research because of its many applications in the
preparation of series of potentially bioactive molecules. In this area, fluorous
chemistry has recently been used in solid phase peptide synthesis,[50–52] nucleotide synthesis[53] and oligosaccharide synthesis.[50,54] The approach has also been
elegantly applied in the synthesis of small organic molecules.[28,55] Because of the
number of researchers working in this area, commercial companies specializing
in reagents and auxiliaries in such processes exist.[56]
   In 2004, Kumar and Montanari introduced a fluorous tagged trivalent
iodonium compound that can be used as a tag for *t*-Boc based solid phase
peptide synthesis by tagging free amines with a perfluoroheptyl (n-C_7F_{15})
group.[52] It is an efficient fluorous tagging reagent that has the potential to help
in the synthesis of both routine and difficult peptide and protein sequences. The

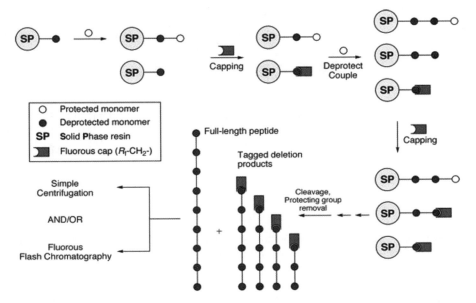

Figure 7.25 Generalized fluorous capping strategy. Amino acids that fail to couple leave an unprotected terminal amino group that is capped with fluorous tagged trivalent iodonium reagent in standard peptide synthesis solvents. All products lacking one (or more) residues are therefore tagged with a fluorous tag that is easily removed at the end of the synthesis by simple centrifugation or by fluorous flash chromatography. [Reprinted with permission from *Eur. J. Org. Chem.*, 2006, 874–877. Copyright 2006 Wiley-VCH.]

fluorous tagging (or capping) process is outlined in Figure 7.25 and has been used in both automated and manual solid phase peptide syntheses.[51] Purification of the peptides can be readily achieved either using centrifugation or by fluorous flash chromatography.

Fluorous reverse phase silica gel (FRPSG) has been used in the purification of synthetic DNA fragments.[53] In solid phase DNA synthesis, truncated sequences are often separated from the desired product after deprotection using HPLC or electrophoresis. In order to perform, parallel syntheses and separations of nucleotides the 'trityl-on' purification procedure was developed, in which a lipophilic support material is used to separate the desired and undesired product, followed by deprotection. If the protecting group is labelled with a fluorous group, fluorous–fluorous interactions between the FRPSG and the protected nucleotide can be used to aid separation of the aqueous mixture.

Oligosaccharide syntheses have also been performed in parallel using a fluorous support.[50,54] A benzyl-type protecting group (HfBn) used in some of these procedures is shown in Figure 7.26. The novel approach here is that the fluorous group or tag is recycled and this can be achieved fairly easily

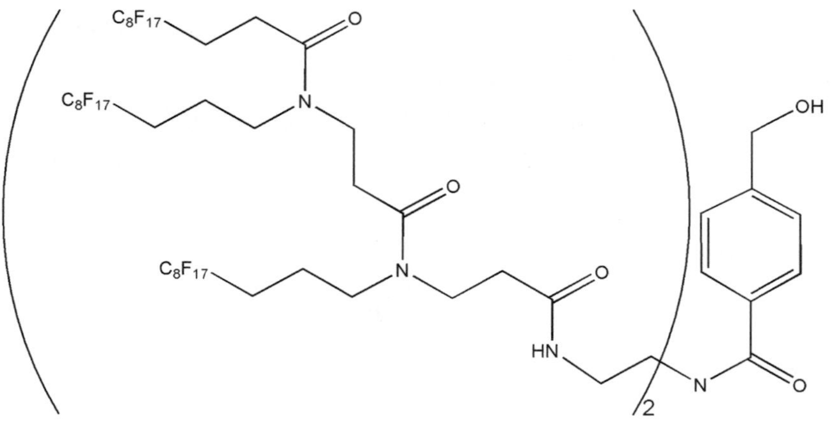

Figure 7.26   Benzyl-type fluorous tag, HfBn(OH), used in oligosaccharide synthesis on a recyclable fluorous support.

by partitioning into a fluorous solvent such as FC72. Pure peptide and oligosaccharide strands can be obtained in high yields, up to 94% even after 13 steps. The use of recyclable fluorous tag, albeit using sacrificial linker units, is a valuable step towards green and sustainable parallel synthesis of biomolecules.

7.2.6 Fluorous Materials Chemistry

Compared to many other alternative solvents, the use of fluorous media in materials chemistry remains significantly underrepresented beyond its use in the preparation of fluorous supports for catalysis and separation. Recently, fluorous labelling, using heptadecafluoro-1-decanethiol, has been shown to be effective in the solubilizing of gold and CdSe nanoparticles in fluorous solvents through phase transfer from an aqueous or hydrocarbon medium (Figure 7.27).[57] Similarly, Rao and co-workers showed that single-walled carbon nanotubes and zinc oxide nanorods can be solubilized in a fluorous medium by reacting them with a fluorous amine, heptadecafluoro-undecylamine. Since the fluorocarbon extracts only the materials containing a fluorous label, the method has potential uses in purifying them. Also, the high non-polarity of fluorous solvents makes it possible to study the optical and other properties of nanostructures in a medium of very low refractive index.

Gold particles have also formed in FBS hydrosilation reactions (see Figure 7.18),[37] and perfluorotagged palladium nanoparticles have been used in carbon–carbon bond-forming reactions.[39] Even more recently, gold nanoparticles with perfluorothiolate ligands have been prepared and studied.[58] Therefore, this field of fluorous chemistry is growing rapidly and holds great potential for future advances in materials chemistry.

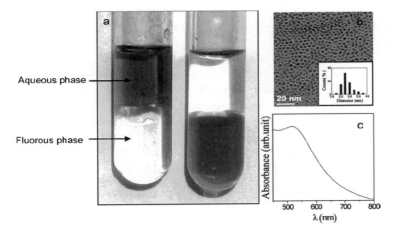

Figure 7.27 (a) Photograph showing transfer of gold nanoparticles (darker phase) from an aqueous medium to the fluorous medium, (b) TEM image and size distribution histogram and (c) UV-Vis absorption spectrum of nanoparticles in fluorous medium. [Reprinted with permission from *J. Phys. Chem. B*, 2006, **110**, 20752–20755. Copyright 2006 The American Chemical Society.]

7.3 Summary and Outlook for the Future

Fluorous technology has been applied to diverse areas of chemistry during the last 15 years. Several very important advances suggest a bright future for fluorous methodologies. Although the methods have not been used on an industrial scale, many fluorous solvents and reagents are now commercially available. Continuous reactors have been developed that allow fluorous biphase catalytic methods to be used without loss of catalyst. This may allow catalytic reactions to be performed homogeneously when the catalyst or method is incompatible with an aqueous biphasic approach. Techniques have been developed to perform fluorous chemistry without the need for large amounts of fluorous solvents by using polymer or fluorous silica supports. These methods, and an increasingly diverse range of fluorous reagents, have led to extensive use of these approaches in medicinal chemistry research using combinatorial or high-throughput methods. Interesting results have recently been obtained in the areas of materials chemistry and nanoparticle syntheses. There are likely to be more advances in this high-value area in the near future, as the additional costs of fluorous media can be outweighed by the opportunities made available by working in a low-polarity medium.

References

1. I. T. Horváth and J. Rábai, *Science*, 1994, **266**, 72.
2. J. A. Gladysz, D. P. Curran and I. T. Horvath, *in Handbook of Fluorous Chemistry*, VCH, Weinheim, 2004.

3. W. Zhang, *QSAR Comb. Sci.*, 2006, **25**, 679.
4. D. P. Curran, K. Mikami and V. A. Soloshonok, *J. Fluor. Chem.*, 2006, **127**, 454.
5. J. A. Gladysz, *Angew. Chem. Int. Edit.*, 2005, **44**, 5766.
6. J. A. Gladysz and D. P. Curran, *Tetrahedron*, 2002, **58**, 3823.
7. L. P. Barthel-Rosa and J. A. Gladysz, *Coord. Chem. Rev.*, 1999, **192**, 587.
8. M. A. Ubeda and R. Dembinski, *J. Chem. Educ.*, 2006, **83**, 84.
9. H. Nakamura, B. Linclau and D. P. Curran, *J. Am. Chem. Soc.*, 2001, **123**, 10119.
10. M. Wende and J. A. Gladysz, *J. Am. Chem. Soc.*, 2003, **125**, 5861.
11. M. Wende, R. Meier and J. A. Gladysz, *J. Am. Chem. Soc.*, 2001, **123**, 11490.
12. B. K. Freed, J. Biesecker and W. J. Middleton, *J. Fluor. Chem.*, 1990, **48**, 63.
13. B. K. Freed and W. J. Middleton, *J. Fluor. Chem.*, 1990, **47**, 219.
14. Y. Marcus, *Chem. Soc. Rev.*, 1993, **22**, 409.
15. Q. Chu, M. S. Yu and D. P. Curran, *Tetrahedron*, 2007, **63**, 9890.
16. M. S. Yu, D. P. Curran and T. Nagashima, *Org. Lett.*, 2005, **7**, 3677.
17. K. L. O'Neal, S. Geib and S. G. Weber, *Anal. Chem.*, 2007, **79**, 3117.
18. K. N. West, J. P. Hallett, R. S. Jones, D. Bush, C. L. Liotta and C. A. Eckert, *Ind. Eng. Chem. Res.*, 2004, **43**, 4827.
19. C. D. Ablan, J. P. Hallett, K. N. West, R. S. Jones, C. A. Eckert, C. L. Liotta and P. G. Jessop, *Chem. Commun.*, 2003, 2972.
20. J. G. Weers, *J. Fluor. Chem.*, 1993, **64**, 73.
21. T. L. Merrigan, E. D. Bates, S. C. Dorman and J. H. Davis, *Chem. Commun.*, 2000, 2051.
22. J. van den Broeke, F. Winter, B. J. Deelman and G. van Koten, *Org. Lett.*, 2002, **4**, 3851.
23. I. T. Horváth, *Acc. Chem. Res.*, 1998, **31**, 641.
24. T. Maruyama, K. Nakashima, F. Kubota and M. Goto, *Anal. Sci.*, 2007, **23**, 763.
25. P. G. Boswell, E. C. Lugert, J. Rabai, E. A. Amin and P. Buhlmann, *J. Am. Chem. Soc.*, 2005, **127**, 16976.
26. P. G. Boswell, C. Szijjarto, M. Jurisch, J. A. Gladysz, J. Rabai and P. Buhimann, *Anal. Chem.*, 2008, **80**, 2084.
27. B. Lemaire, B. Mompon, I. Surbled and M. Surbled, in *Method for Fractionating Essential Oils Using at Least a Fluorinated Solvent*, Patent Application 10/312,223, U. S. Patent Office Office, Washington, DC, 2004.
28. W. Zhang, Z. Y. Luo, C. H. T. Chen and D. P. Curran, *J. Am. Chem. Soc.*, 2002, **124**, 10443.
29. K. E. Myers and K. Kumar, *J. Am. Chem. Soc.*, 2000, **122**, 12025.
30. Q. L. Chu, W. Zhang and D. P. Curran, *Tetrahedron Lett.*, 2006, **47**, 9287.
31. S. Dandapani and D. P. Curran, *Tetrahedron*, 2002, **58**, 3855.
32. D. P. Curran, *Aldrichimica Acta*, 2006, **39**, 3.
33. J. Horn and W. Bannwarth, *Eur. J. Org. Chem.*, 2007, 2058.
34. R. C. da Costa and J. A. Gladysz, *Adv. Synth. Catal.*, 2007, **349**, 243.

35. J. Otera, *Acc. Chem. Res.*, 2004, **37**, 288.
36. L. V. Dinh and J. A. Gladysz, *New J. Chem.*, 2005, **29**, 173.
37. D. Lantos, M. Contel, A. Larrea, D. Szabo and I. T. Horváth, *QSAR Comb. Sci.*, 2006, **25**, 719.
38. E. G. Hope, J. Sherrington and A. M. Stuart, *Adv. Synth. Catal.*, 2006, **348**, 1635.
39. R. Bernini, S. Cacchi, G. Fabrizi, G. Forte, S. Niembro, F. Petrucci, R. Pleixats, A. Prastaro, R. M. Sebastian, R. Soler, M. Tristany and A. Vallribera, *Org. Lett.*, 2008, **10**, 561.
40. C. C. Tzschucke, V. Andrushko and W. Bannwarth, *Eur. J. Org. Chem.*, 2005, 5248.
41. C. C. Tzschucke, C. Markert, H. Glatz and W. Bannwarth, *Angew. Chem. Int. Edit.*, 2002, **41**, 4500.
42. W. B. Yi, C. Cai and X. Wang, *Eur. J. Org. Chem.*, 2007, **1**, 3445.
43. A. Gheorghe, E. Cuevas-Yañez, J. Horn, W. Bannwarth, B. Narsaiah and O. Reiser, *Synlett*, 2006, **17**, 2767.
44. C. S. Consorti, M. Jurisch and J. A. Gladysz, *Org. Lett.*, 2007, **9**, 2309.
45. A. Yoshida, X. Hao, O. Yamazaki and J. Nishikido, *QSAR Comb. Sci.*, 2006, **25**, 697.
46. A. Yoshida, X. H. Hao and J. Nishikido, *Green Chem.*, 2003, **5**, 554.
47. E. Perperi, Y. L. Huang, P. Angeli, G. Manos, C. R. Mathison, D. J. Cole-Hamilton, D. J. Adams and E. G. Hope, *Dalton Trans.*, 2004, 2062.
48. H. R. Hobbs, H. M. Kirke, M. Poliakoff and N. R. Thomas, *Angew. Chem. Int. Ed.*, 2007, **46**, 7860.
49. J. L. Panza, A. J. Russell and E. J. Beckman, *Chem. Commun.*, 2002, 928.
50. M. Mizuno, K. Goto, T. Miura and T. Inazu, *QSAR Comb. Sci.*, 2006, **25**, 742.
51. V. Montanari and K. Kumar, *Eur. J. Org. Chem.*, 2006, 874.
52. V. Montanari and K. Kumar, *J. Am. Chem. Soc.*, 2004, **126**, 9528.
53. C. Beller and W. Bannwarth, *Helv. Chim. Acta*, 2005, **88**, 171.
54. K. Goto, T. Miura, M. Mizuno, H. Takaki, N. Imai, Y. Murakami and T. Inazu, *Synlett*, 2004, 2221.
55. W. Zhang and Y. Lu, *J. Comb. Chem.*, 2007, **9**, 836.
56. Fluorous Technologies Inc., *www.fluorous.com*, accessed June 2008.
57. R. Voggu, K. Biswas, A. Govindaraj and C. N. R. Rao, *J. Phys. Chem. B*, 2006, **110**, 20752.
58. A. Dass, R. Guo, J. B. Tracy, R. Balasubramanian, A. D. Douglas and R. W. Murray, *Langmuir*, 2008, **24**, 310.

CHAPTER 8
Liquid Polymers

8.1 Introduction

Low molecular weight polymers or those with low glass transition temperatures can be used as non-volatile solvents. In particular, poly(ethyleneglycols) (PEGs) and poly(propyleneglycols) (PPGs) have been used recently in a range of applications.[1] Just like fluorous solvents, these can be used to enable recycling of valuable catalysts and other chemicals. They show interesting phase behaviour and allow the benefits of a heterogeneous and homogeneous system to be employed by adjusting an external variable such as temperature.

PEGs are available in a wide range of molecular weights, and complete toxicity profiles are available. They are components in many consumer products such as shampoos and other personal care items, and have been approved by the U.S. Food and Drug Agency for internal consumption. PEGs and PPGs are water soluble and therefore many of their applications involve aqueous solutions. (The higher the molecular weight of the polymer, the lower its solubility in aqueous solution. PPG is less hydrophilic than PEG of a comparable molecular weight.) PEG has low flammability and low (to zero) vapour pressure. In contrast to many alternative solvents, PEG is known to be biodegradable, biocompatible and therefore non-toxic. PEG can be recovered and recycled from solutions by extraction or direct distillation of the volatile component.

8.1.1 Properties of Aqueous PEG Solutions

Although PEG is water soluble, upon varying the temperature of a solution it can form distinct polymer-rich and polymer-poor phases. This is due to the hydrophobic methylene groups along the backbone of the polymer (Figure 8.1), interspersed with the hydrophilic ether groups and alcohol end groups. This

RSC Green Chemistry Book Series
Alternative Solvents for Green Chemistry
By Francesca M. Kerton
© Francesca M. Kerton 2009
Published by the Royal Society of Chemistry, www.rsc.org

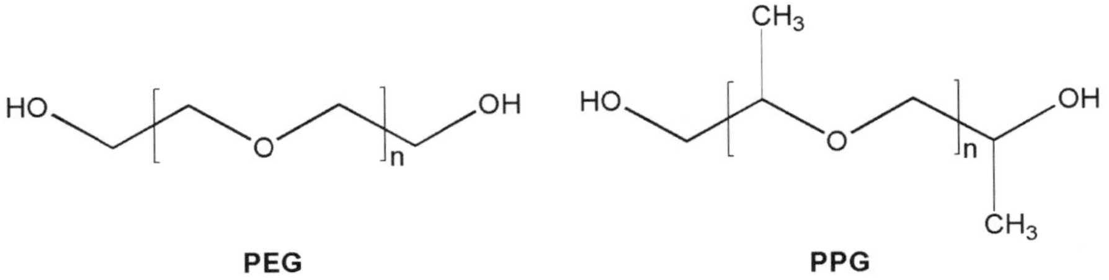

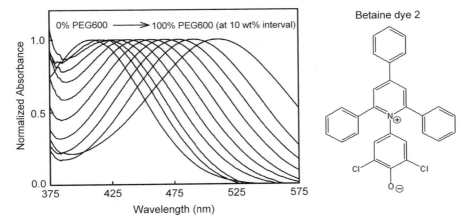

Figure 8.1 General structures of PEG and PPG.

Figure 8.2 Normalized UV-Vis absorbance spectra for PEG 600–water mixtures of varying compositions containing betaine dye 2. [Reprinted with permission from *Green Chem.*, 2007, **9**, 254–261. Copyright 2007 The Royal Society of Chemistry.]

phase behaviour is also affected by the presence of salts, *e.g.* sodium hydrogen sulfate or potassium phosphate, and results in the formation of aqueous biphasic systems (ABS). Phase separations such as these have been exploited in bio-separations for some time and are now being used in chemical applications.[1]

In chemical applications, the PEG acts as a co-solvent and imparts an apparent decrease in solution polarity, which leads to an increase in solubility of organic molecules. Solution polarity measurements made using a water soluble betaine dye (Figure 8.2) afford spectra that exhibit the expected bathochromic shift in λ_{max} as the wt% of PEG is increased. These data suggest that a decrease in polarizability and/or hydrogen bond donating (HBD) acidity is occurring as PEG is added to water (Figure 8.3).[2] Measurements have also been made on PEGs in buffered aqueous solutions.[1]

Sometimes it is useful to have a rule of thumb as to which conventional solvent to compare an alternative with: for aqueous PEG solutions $E_T(30)$ or $E_T(33)$ values are typically between 55 and 70, indicating a polarity similar to

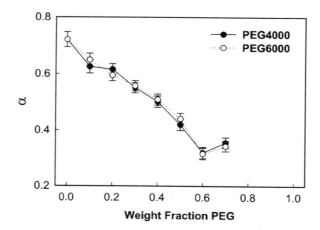

Figure 8.3 Hydrogen bond donating acidity (α) *vs* weight fraction of PEG in aqueous PEG for PEG 4000 ($\bullet$) and PEG 6000 ($\bigcirc$). [Reprinted with permission from *Green Chem.*, 2007, **9**, 254–261. Copyright 2007 The Royal Society of Chemistry.]

short chain alcohols such as methanol and propanol. The presence of an organic substance (a third component) will decrease this value. Further details on solvation models and relationships for ABS can be found in the review by Rogers and co-workers.[1] As hydrophilic polymers, PEG and low molecular weight PPG are soluble in water; however, they are also soluble in many organic solvents including toluene and dichloromethane. They are insoluble in aliphatic hydrocarbons, which could be used to extract compounds from the polymer phase. Liquid PEGs and PPGs can also be thought of as protic solvents with aprotic sites of binding at each ethylene–propylene oxide monomer unit. PPG is more viscous and less soluble in water than PEG, and this may have led to less exploration of its use as a replacement reaction medium to date. As it is more hydrophobic than PEG, this may lead to some interesting alternative uses for PPG as a relatively unexploited alternative reaction medium. Another class of liquid polymers is the poly(dimethylsiloxane)s. These are very hydrophobic and are just starting to be used as non-volatile reaction media.[3] Other polymers of low molecular weight may also be liquids and may hold potential as solvents.

PEGs have been extensively used to date in aqueous biphasic systems (ABS). Therefore, it is important to understand their phase behaviour, although this involves many variables (polymer molecular weight, salts, neutral organic molecules, temperature, *etc.*), it will ultimately lead to a better understanding of chemistry in these alternative solvents. Indeed, Rogers and co-workers have already shown that the distribution of organic solutes in these systems is a function of the difference in polymer concentration between the polymer-rich and polymer-poor (aqueous) phases.[4] Also, a series of near identical ABSs can be prepared even when the salts used are different (they will just possess a

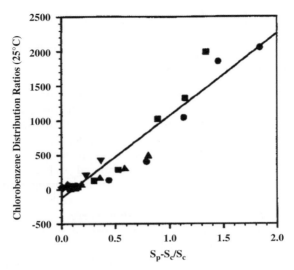

Figure 8.4 Distribution ratios for chlorobenzene *vs* the reduced salt concentration [SR] in ABSs (aqueous biphasic systems) formed with 40% (w/w) PEG 2000 and increasing concentrations of salt: ● K_3PO_4, ■ K_2CO_3, ▲$(NH_4)_2SO_4$, ▼ NaOH. [Reprinted with permission from *Ind. Eng. Chem. Res.*, 2002, **41**, 1892–1904. Copyright 2002 American Chemical Society.]

different ratio of PEG to salt).[4] This perhaps somewhat simplifies the behaviour of PEG-ABS and leads to a situation where data for all salts used can be seen to follow a general trend line for the distribution of an organic molecule between the two phases (Figure 8.4).

As PEG has many ether groups along the length of its polymer chain, it is able to form metal ion complexes similar to crown ether metal complexes. This is one of the reasons why salts have such a significant effect on the phase behaviour of PEG in aqueous solutions. These metal complexes can then be used as phase transfer catalysts (PTCs). Considering the significant differences in price between PEGs and crown ethers, as well as their lower toxicity, such species are likely to find significant applications in the future.[1] Their catalytic behaviour is dependent on the PEG molecular weight, any end group modifications and the nature of the cation (*e.g.* Na^+, Ln^{3+}) and the anion (*e.g.* OH^-, HCO_3^-, Cl^-, NO_3^-). PEGs are stable to oxidizing conditions and have been used in organic chemistry with stoichiometric oxidizing agents, *e.g.* $K_2Cr_2O_7$.

8.2 Chemical Examples

8.2.1 PEG and PPG as Non-volatile Reaction Media

A wide range of reactions and chemical processes have used PEG or PPG. The liquid polymer can be used directly as an inert, non-volatile solvent, or it can be modified and used as a catalyst or reactant molecule. PEGs have been used widely in ABS and PTC.

8.2.1.1 PEG as a Reaction Solvent

Liquid polyethylene glycol in organic synthesis. Three main types of reaction have been studied: substitution, oxidation and reduction (Figure 8.5).[1] The Diels–Alder reaction is frequently used as a benchmark for alternative reaction media. For the reaction of 2,3-dimethyl-1,3-butadiene with nitrosobenzene in PEG 300 or PPG 425, a 3.3-fold increase in rate was seen compared with dichloromethane and a 2.5-fold increase compared with ethanol.[5] It has also been shown more recently that in comparison to conventional solvents, the transition state for the Diels–Alder reaction is stabilized in aqueous PEG solutions and this results in a lower activation energy.[6]

Another atom-efficient reaction that has been often studied in alternative solvents is the Michael addition. Chandra and co-workers have shown that this proceeds most efficiently in PEG and in the absence of added catalyst (Figure 8.6).[7] Reaction times are dramatically shorter than in conventional solvents and give near quantitative yields of the addition product. It was proposed that the hydroxyl end groups of the polymer act to weaken the N–H bond of the amine and this increases the nucleophilicity of the nitrogen for addition to the alkene. For other Michael addition reactions in PEG, added catalysts have been used. For example, 0.5 mol% RuCl$_3$ in PEG was successfully employed in aza-Michael additions of α,β-unsaturated carbonyls with aliphatic and aromatic amines, and also with thiols and carbamates.[8] Excellent yields are obtained and the regioselectivity for the reaction is improved compared with dichloromethane. The catalyst-containing PEG phase could be recycled five times. More recently, Michael additions in PEG have been used for the synthesis of N3-functionalized dihydropyrimidines using potassium carbonate as the catalyst.[9] Organocatalysed asymmetric Michael addition reactions have also been performed in PEG, where it was proposed that a PEG–organocatalyst host–guest complex forms.[10] Enantioselectivities were generally much higher in the PEG systems than when the same reactions were performed using the organocatalyst in a conventional solvent such as DMSO or THF.

Because of the low volatility of these liquid polymer solvents, they have often been used in conjunction with microwave heating, which can dramatically reduce reaction times. Microwave assisted palladium catalysed Suzuki cross coupling of arylboronic acids and aryl halides has been performed in PEG 400 (Figure 8.5).[11] Moderate to excellent yields were obtained and the catalyst-containing Pd–PEG phase could be recycled three times after extracting the product with diethyl ether. The reaction time was only 60 s compared with 15 min heating using an oil bath at 100 °C to achieve the same yields. More recently, Suzuki couplings using between 0.0001 mol% and 3 mol% palladium have been performed in PEG 400 using DABCO as the base and TBAB as a promoter.[12] The catalyst system could be recycled and reused five times without

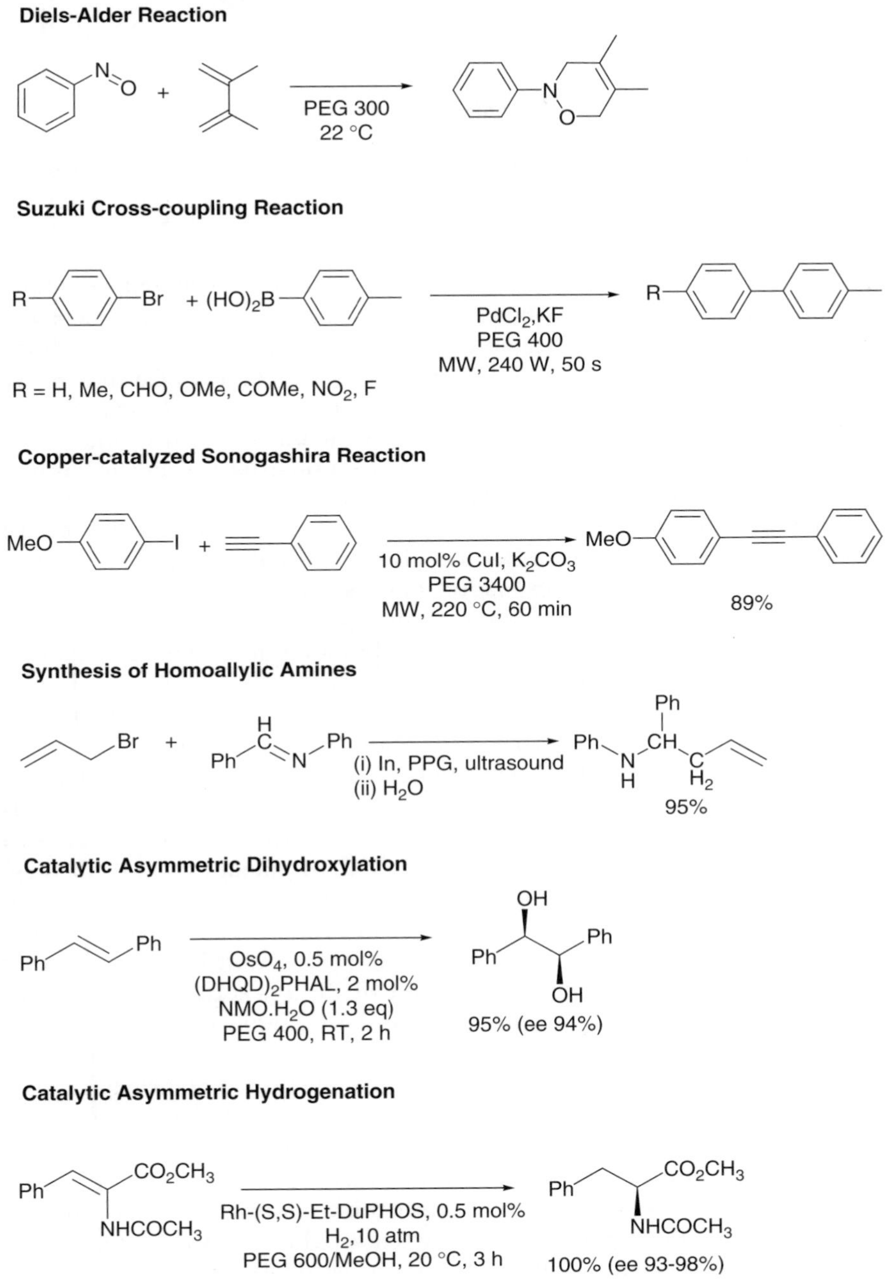

Diels-Alder Reaction

Suzuki Cross-coupling Reaction

R = H, Me, CHO, OMe, COMe, NO_2, F

Copper-catalyzed Sonogashira Reaction

89%

Synthesis of Homoallylic Amines

95%

Catalytic Asymmetric Dihydroxylation

95% (ee 94%)

Catalytic Asymmetric Hydrogenation

100% (ee 93-98%)

Figure 8.5 Some synthetic organic reactions performed in PEG and PPG.

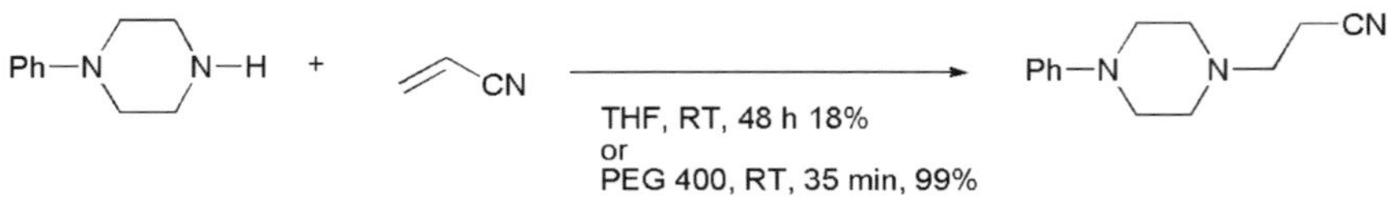

THF, RT, 48 h 18%
or
PEG 400, RT, 35 min, 99%

Figure 8.6   Example of a Michael addition reaction of an amine to a conjugated alkene.

any loss in activity, and TONs using this system were as high as 960 000! Other types of palladium catalysed coupling reactions, *e.g.* Heck reactions, have also been performed in PEGs.[13] Cheaper copper based catalyst systems for carbon–carbon coupling in PEG have also been developed. The use of a microwave allows this less reactive metal to be used, and it has been suggested that the PEG solvent plays a pivotal role in preventing undesired side reactions.[14,15]

PPG 425 has been used in the preparation of a range of benzaldehydes that were subsequently used in the solvent free synthesis of calix[4]resorcinarenes.[16] The aldehydes are isolated in near quantitative yields by distillation from the PPG, which can be recycled. PPG has also been used in the indium metal mediated allylation of imines and sulfonylimines. Ultrasound was used for two reasons: to clean the metal surface and to increase the solubility of the imine in PPG. The solvent was recycled three times, but a desire to develop an easier drying and recycling method was indicated.[17]

Schiff base condensation and modified Mannich condensation reactions have been performed in PPG and PEG to yield a wide variety of ligand molecules in high yields.[18,19] The polymer solvent can be recycled several times but superior yields are obtained in solvent free syntheses for Schiff bases and in aqueous suspensions for the Mannich procedure. 4′-Pyridyl terpyridines have also been performed using condensation reactions in PEG, although the yields of these products are in the 45–55% range. In contrast, the three-component one-pot Biginelli reaction, which also involves a condensation step, proceeds exceptionally well in the presence of PEG 400.[20] The PEG is described as a promoter of this reaction and an insufficient amount of PEG is used for it to be the solvent. Another three-component reaction, the Passerini reaction, has been performed in both PEG 400 and ionic liquids.[21] The reaction combines a carboxylic acid with an aldehyde and an isocyanide to give α-acyloxy carboxamides which are of potential pharmaceutical interest. Yields were 5–20% higher in PEG 400 than in [Bmim]PF$_6$, and reaction times were shorter: 6 h in PEG compared with 14 h in [Bmim]PF$_6$ and 3 days in THF.

Several oxidation reactions have been performed in PEG.[1] In this book, we will focus on catalytic oxidation reactions. The H$_5$PV$_2$Mo$_{10}$O$_{40}$ polyoxometalate was very effective in a range of aerobic oxidation reactions in PEG 200 (Table 8.1).[22] The solvent–catalyst phase could be recovered and reused. High-yielding and selective Sharpless-type asymmetric dihydroxylations can be achieved rapidly using smaller than normal amounts of toxic osmium tetroxide (0.5 mol% *vs* 1–5 mol%) (Figure 8.5).[23] The expensive asymmetric ligand and

Table 8.1   Some catalytic oxidations using $H_5PV_2Mo_{10}O_{40}$ in PEG 200.[22]

Reaction	Substrate	Product (selectivity/%)
Oxydehydrogenation of alcohols	Benzyl alcohol	Benzaldehyde (100)
	4-Bromobenzyl alcohol	4-Bromobenzaldehyde (100)
	1-Phenyl ethanol	Styrene (66), acetophenone (26), benzaldehyde (8)
Oxydehydrogenation of dienes	α-Terpinene	4-Cymene (100)
	Limonene	4-Cymene (80), γ-terpinene (11), α-terpinolene (9)
	4-Vinylcyclohexene	Ethylbenzene (100)
Oxidation of sulfides	Tetrahydrothiophene	Sulfoxide (73), sulfone (27)
	Dibutyl sulfide	Sulfoxide (61), sulfone (39)
	Thioanisole	Sulfoxide (77), sulfone (23)
Wacker reaction	Propene	Acetone (100)

osmium tetroxide remain in the PEG phase and can be recycled four times with no significant drop in activity. The product phase, produced by ether extraction, contains less than 2 ppm osmium.

As a counterpoint to oxidation reactions, it is worth noting that several catalytic reduction reactions of C=C and C=O bonds have also used PEG as a reaction medium. For example, PEG 600 has been used as a recyclable catalyst-containing phase in enantioselective hydrogenations of 2-arylacrylic acids, β-keto esters, aryl ketones and enamides using ruthenium or rhodium catalysts (Figure 8.5).[24] Improved yields and enantioselectivities were obtained when methanol was used as a co-solvent, but the PEG was essential to aid in recycling the expensive catalyst. A recycling study in the hydrogenation of enamides using a Rh-DuPHOS catalyst showed that nine reactions could be performed without any significant drop in conversions (>99%) or enantioselectivity (ee >93%). It has since been shown that water can be used as the co-solvent to give recyclable ruthenium catalysts in PEG 400 for enantioselective transfer hydrogenations.[25] It should also be noted that in some examples the hydroxyl end groups on PEG can inhibit catalysis. This has been overcome for some iridium hydrogenation catalysts by using poly(ethylene glycol) dimethyl ether (DMPEG),[26] where the –OH groups have been replaced by –OMe groups. Products could still be easily extracted from the reaction mixture using a non-polar solvent, *e.g.* hexane, and the iridium catalyst remained immobilized in the DMPEG and was recycled seven times.

A range of solvents have been used to extract products from reactions using PEGs; most notably, they have been used with scCO$_2$. Heldebrant and Jessop used PEG 900 in the hydrogenation of styrene using RhCl(PPh$_3$)$_3$ as the catalyst.[27] The styrene, PEG and Rh catalyst were heated to 40 °C under 30 bar hydrogen and 50 bar carbon dioxide in a stainless steel pressure vessel. After 19 h, the temperature was increased to 55 °C and scCO$_2$ was bubbled through

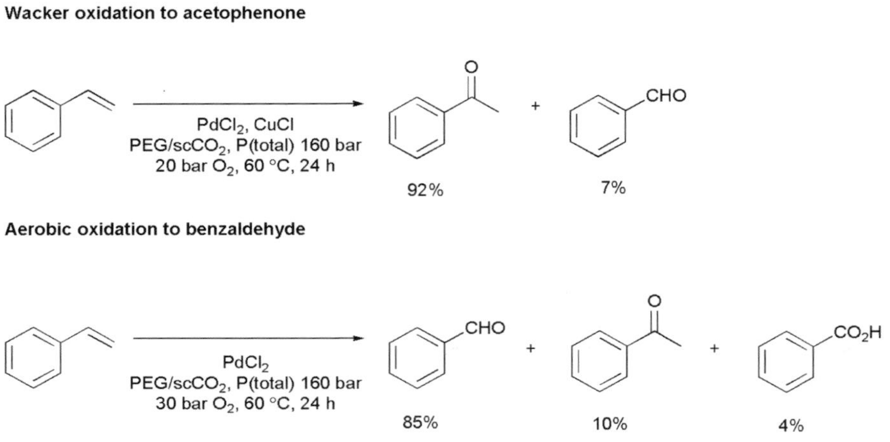

Wacker oxidation to acetophenone

92%   7%

Aerobic oxidation to benzaldehyde

85%   10%   4%

Figure 8.7 Palladium-catalysed oxidation of styrene in PEG–scCO$_2$.

the PEG phase at 155 bar and 2 mL min^{-1}. The released gaseous phase was collected in a cooled trap of dichloromethane solvent. The contents of the reaction vessel (PEG phase) was recycled five times without the need to add more catalyst or PEG. The catalyst remained active (>99% conversion), the rhodium content of the product (ethylbenzene) was below detectable levels and very little PEG was found to leave the reaction vessel.

A PEG–scCO$_2$ system has also been used in the aerobic oxidation of styrene (Figure 8.7).[28] In the presence of cuprous chloride co-catalyst the reaction favours acetophenone formation, whereas in the absence of copper benzaldehyde is favoured. The catalyst could be recycled five times and it was suggested that the PEG acts to prevent the palladium catalyst from decomposing and also assists in product separation.

PEG and PEG-derived complexes and compounds have been intensively investigated in phase transfer catalysis, often because of their high thermal stability, low cost and ability to form crown ether like complexes. An overview of their use in this area is given by Rogers.[1] They have been used in Williamson ether synthesis with or without organic solvents, and yields are generally comparable to PTC systems utilizing expensive and potentially toxic crown ethers or cryptands. As in their use in reaction media, PEGs have been used in PTC nucleophilic substitution reactions, oxidation and reduction reactions. Oxidations include alcohols to aldehyde, benzyl halides to esters and acids, and styrenes to acids. Reductions include ketones to secondary alcohols, and aldehydes or esters to alcohols. PEGs have also been used as polymeric supports for other sorts of PTC, and further details can be found in the review by Janda and co-workers.[29] In many cases, the supported reagent or catalyst can be used as the solvent as well as the support. For example, PEG 3400 has been used in the microwave assisted parallel synthesis of amino acid derivatives and shows some advantages over a solid phase synthetic route.[30] Some recent

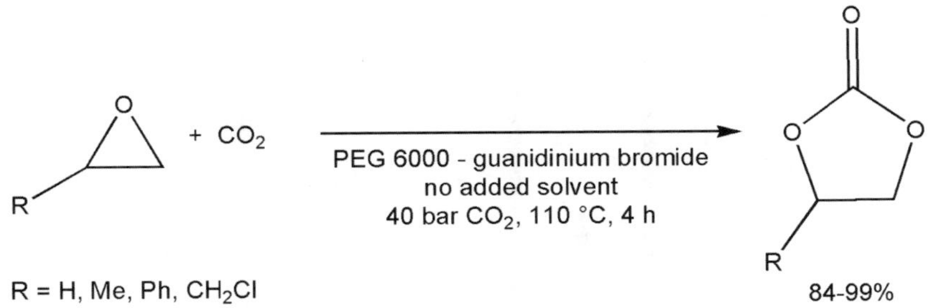

R = H, Me, Ph, CH₂Cl

84-99%

Figure 8.8   Functionalized PEG in 'solventless' catalytic synthesis of cyclic carbonates from carbon dioxide and epoxides.

results of particular interest to green chemists are the use of guanidinium salt functionalized PEG and phosphonium halide functionalized PEG in catalytic carbon dioxide fixation (Figure 8.8).[31,32] A carbon dioxide pressure of 40 bar was needed to achieve quantitative conversion to the carbonate in the guanidinium-PEG procedure, whereas less than 2 bar was required for the phosphonium-PEG catalyst and the catalyst could be reused five times with no loss in activity.

Polyethylene glycol in aqueous biphasic reactive extraction.   Aqueous biphasic reactive extraction (ABRE) is used to describe the use of ABS in biphasic reaction chemistry. When it is used effectively, it can aid in separating reactants, products and catalysts, and increase yields and selectivities. Although, PEG and ABS have been used in bio-separations since the 1950s, it is only during the last 10 years that they have been explored as green reaction media.[1,33] The three main characteristics offered by ABRE in this area are:[1]

1. Phase separation of reactants and products that can act to drive the reaction forward.
2. A PEG-rich top phase in PEG-salt ABS, which has organic solvent type properties and can act as a reaction medium.
3. Catalytic PEG and metal complexes can be used and separated after the reaction.

At this point, it is worth pointing out that the phase behaviour in such systems can be quite complicated and reactions can proceed in a three phase (or even higher) manner. A good example of this is the triphase synthesis of *n*-butyl phenyl ether from sodium phenolate and *n*-butyl bromide using PEG 600 as the catalyst.[34] Sodium hydroxide was found to be the most effective salting-out agent and non-polar heptane the most effective organic solvent. The choice of the salt in obtaining good phase separation and reaction rates is essential and needs identifying for each new reaction attempted. In the catalytic oxidation of cyclic olefins using aqueous hydrogen peroxide as the oxidant, sodium

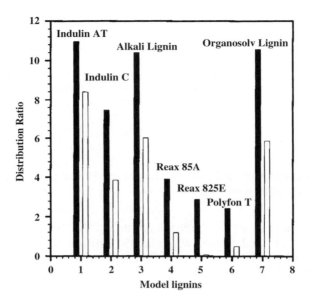

Figure 8.9 Comparison of distribution of model lignins in PEG 2000–pulping solution ABS (■) and PEG 2000/NaOH ABS (□). [Reprinted with permission from *Ind. Eng. Chem. Res.*, 2002, *41*, 2535–42. Copyright 2002 American Chemical Society.]

hydrogen sulfate was the preferred salt.[33] Multiphase systems and alternative solvents in general will continue to be of significant interest in the field of green oxidation chemistry so as to control these important reactions and prevent competing side reactions.

In biphasic reaction chemistry, exciting results have been achieved in the area of wood delignification (of relevance to the paper industry).[1,35,36] As shown in Figure 8.9, the type of salts present in the formation of a multiphase PEG based system have a practical effect on the distribution ratios and partitioning of lignins. Pulping solutions contain sodium sulfite and sodium carbonate in addition to the strong base sodium hydroxide. With all types of lignins studied, the authors found that the presence of these additional salts had a positive effect on the phase separation. In summary, ABRE has shown the following advantages in this area:[1,35,36]

1. No organic solvent is required. (An alcohol is used in organosolvent pulping.)
2. Salts added to aid phase separation, *e.g.* lithium sulfate, can act as catalysts in the delignification process.
3. The reaction is enhanced as a result of opposite partitioning of the cellulose and lignin components.
4. Fibres are swollen. This improves access for reagents and increases the reaction rate.

In addition to the application of ABRE in wood chemistry, it has been widely applied to other biomass conversion fields, particularly enzyme catalysed hydrolysis reactions such as the conversion of biopolymers (including cellulose and starch) to monosaccharides and oligosaccharides.[1] In this area, PEG ABS systems offer a benign non-denaturing environment in contrast to organic solvent reaction media.

Polyethylene glycol in the synthesis of materials. PEG has been used as a solvent in polymerization reactions. It was found to facilitate easy removal of the metal catalyst in transition metal mediated living radical polymerization (Figure 8.10).[37] Products from this type of polymerization are usually heavily contaminated with intensely coloured copper impurities. In the case of methyl methacrylate polymerization the reaction rate was higher than in conventional organic solvents, but for styrene the reaction was slower than in xylene.

Recently, a new method for the preparation of silver coatings has been developed by Tanemura and co-workers.[38] This technique may find applications in the electronics industry. When a solution of silver oxide in PEG 400 or PEG 500 dimethyl ether is heated for 8 h, a film forms on the walls of the flask (Figure 8.11). In contrast, when other silver salts are treated in the same way they yield precipitates. As well as utilizing a safe solvent, this procedure is also interesting as no reducing agent is needed. Autoxidation of the solvent occurs and this results in the reduction of silver oxide. During the course of this study it was found that copper(I) oxide, copper(II) oxide, tin(II) oxide, cobalt oxide (Co_3O_4), gold(III) oxide and titanium(III) oxide were also appreciably soluble in PEG400, but thin metal coatings could not be formed using the same method.

In addition to films of silver, bulk syntheses of silver and iron nanorods can be performed in PEG (Figure 8.12).[39] The formation of these particles takes just a few minutes in a microwave reactor. However, to get uniform particle

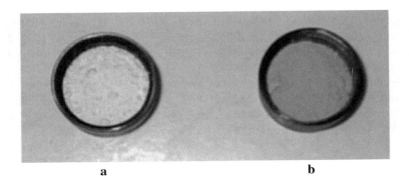

a   b

Figure 8.10 Precipitated PMMA from reaction in (a) PEG 400 and (b) toluene, without filtration through basic alumina. [Reprinted with permission from *Chem. Commun.*, 2004, 604–605. Copyright 2004 The Royal Society of Chemistry.]

Figure 8.11   Silver coatings prepared from PEG 400: (a) silver mirror on inside surface of flask, (b) SEM of silver particles after 3 h and (c) after 8 h, (d)
cross-sectional SEM view of film on glass. [Reprinted with permission
from *Chem. Lett.*, 2007, **36**, 782–783. Copyright 2007 The Chemical
Society of Japan.]

morphology longer reaction times are preferred. As in the work with silver
films, no additional reducing agent is needed and therefore there is enormous
potential for producing a range of interesting metal-containing materials in
PEG solution.

Catalytic metal nanoparticles can also be prepared in PEG. Heating a solution
of palladium acetate in PEG (molecular weights from 400 to 4000) at 90 °C for
20 min or longer results in the oxidation of PEG and the reduction of the metal
to yield nanoparticles (Figure 8.13).[40] The resulting palladium–PEG catalyst
systems exhibit high activity, selectivity and stability in the hydrogenation of a
wide range of olefins and could be recycled 10 times with no loss in reactivity.
Similar palladium nanoparticles have been observed using TEM in the recyclable PEG phase from Sonogashira carbon–carbon coupling reactions.[41]

8.2.2   Poly(dimethylsiloxane) as a Non-volatile Reaction Medium

To date there have been very few investigations into non-volatile liquid polymers other than PEG used in chemical reactions. PPG has been used to some

Figure 8.12 SEM images of Ag nanorods synthesized using microwave irradiation: (a) and (b) bulk morphologies are nanorods, (c) and (d) from mixtures more concentrated in silver nitrate contain nanoparticles in addition to nanorods. [Reprinted with permission from *Cryst. Growth Des.,* 2008, *8,* 291–295. Copyright 2008 American Chemical Society.]

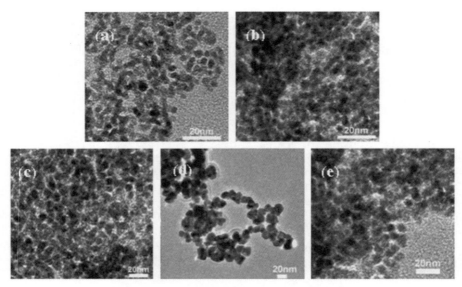

Figure 8.13   TEM photographs of palladium nanoparticles in PEGs: (a) palladium in
              PEG 800, (b) palladium in PEG 1000, (c) palladium in PEG 2000, (d)
              palladium PEG 4000 and (e) palladium PEG 2000 after 10 recycles.
              [Reprinted with permission from *Catal. Commun.*, 2008, **9**, 70–74.
              Copyright 2008 Elsevier B.V.]

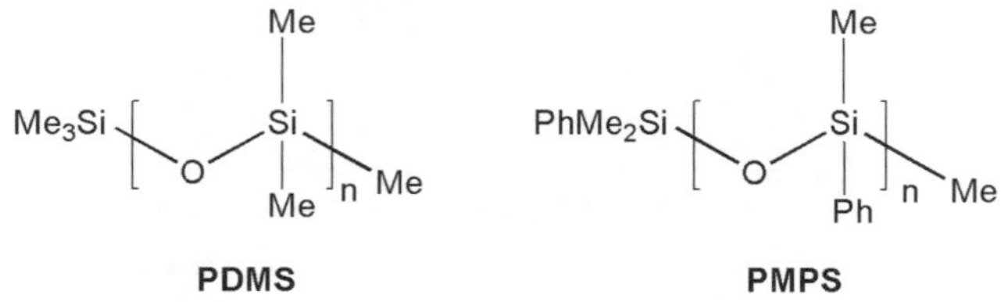

Figure 8.14   General structures of PDMS and PMPS.

extent and has been mentioned in the PEG discussions earlier in this chapter. In
an extension to his work using PEG–scCO$_2$ in catalytic reductions,[27] Jessop
performed a comparative study using different liquid polymers including
poly(dimethylsiloxane) (PDMS) and poly(methylphenylsiloxane) (PMPS)
(Figure 8.14).[3] Whereas PEG and PPG usually possess –OH end groups, these
silicone based materials are usually prepared in the presence of capping agents
such as hexamethyldisiloxane and therefore contain non-protic end groups.
However, end group and side chain functional siloxane materials are also
commercially available.

   PDMS and PMPS possess low toxicities and are widely used in consumer
products, but there are conflicting reports on their biodegradation. UV
absorbance measurements show that PMPS has a polarity intermediate

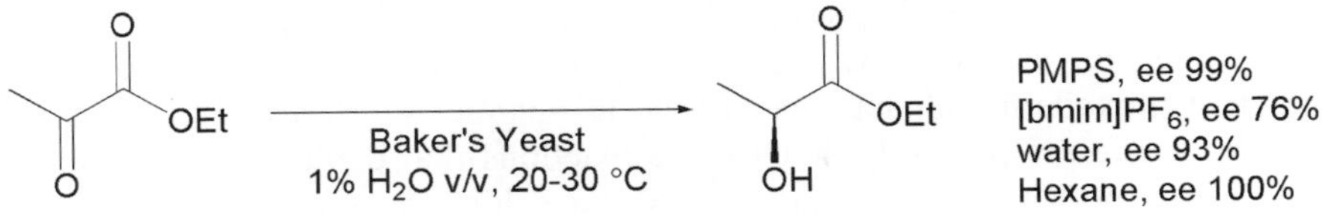

Figure 8.15 Yeast catalysed reductions of ethyl pyruvate.

between toluene and hexane.[3] When ruthenium catalysed asymmetric hydrogenation of tiglic acid was performed in PDMS moderate enantioselectivities were achieved, which was inferior to the results obtained in methanol and [bmim]PF$_6$. Interestingly, whole-cell catalysed reductions could be performed in PMPS and exceptional enantioselectivities were achieved (ee 99%) (Figure 8.15). In this regard, the siloxane polymer was the superior liquid polymer solvent for this reaction and the product could be extracted into water. Therefore, there is promise for PDMS and PMPS in reactions requiring a hydrophobic polymer where the product can be extracted into aqueous media.

PDMS has also been used as a reaction medium in the preparation of polymeric dispersions.[42,43] In these examples, an epoxy end group functionalized PDMS was used and an oil-in-oil emulsion was formed with the reacting monomer (vinyl acetate or vinyl pyrrolidone). There are likely to be further advances in this area during the next few years.

8.3 Summary and Outlook for the Future

PEGs with molecular weights from 200 to 6000 are the polymers most widely used as alternative solvents. PEG is cheap, non-toxic, biodegradable and, because of the large number of ether groups along its backbone, it can form complexes with metal salts that can then be used directly as catalysts or in PTC. At higher molecular weights, despite being a solid, it can either be heated to give a liquid reaction medium or used in conjunction with water in ABS. Given the extensive research carried out using PEG during the last 10 years, there is now a better level of understanding of its properties as a reaction medium and in some cases it gives superior results to conventional solvents and other alternative reaction media, *e.g.* ionic liquids. In contrast, surprisingly little chemistry has been performed in its more hydrophobic relation, liquid PPG. New catalysts are being developed for the stereospecific ring-opening polymerization of propylene oxide that yields PPG with controlled chirality at each methine carbon and therefore one can envisage chiral PPG being used to induce stereoselectivity in a similar way to that recently achieved in chiral ionic liquid media. Also, further work will no doubt continue using functional PEG and PPG in polymer supported syntheses and in recyclable catalyst procedures. The benign nature of aqueous PEG solutions makes them ideal media for biocatalytic reactions and advances in this area are likely to be significant in the next 10 years.

Hydrophobic siloxane based liquid polymers have recently been used as inert reaction media in transition metal catalysed and yeast catalysed reactions. Reactive siloxane liquids have been used in polymer synthesis as the *de facto* solvent. There are probably many more reactions that could be performed in these liquid polymers, and isolation of water soluble products could easily be achieved in a separate aqueous phase.

There are many more classes of polymer in addition to those already studied, and by controlling the molecular weight in their synthesis, liquid polymers can be obtained for nearly all linear homopolymers and copolymers known. Therefore, there is the opportunity to tailor liquid polymer solvents in a similar way to tailoring ionic liquids, and this would lead to a wealth of new chemical applications for these materials, including their use as solvents.

References

1. J. Chen, S. K. Spear, J. G. Huddleston and R. D. Rogers, *Green Chem.*, 2005, **7**, 64.
2. P. Singh and S. Pandey, *Green Chem.*, 2007, **9**, 254.
3. D. J. Heldebrant, H. N. Witt, S. M. Walsh, T. Ellis, J. Rauscher and P. G. Jessop, *Green Chem.*, 2006, **8**, 807.
4. H. D. Willauer, J. G. Huddleston and R. D. Rogers, *Ind. Eng. Chem. Res.*, 2002, **41**, 1892.
5. N. F. Leininger, R. Clontz, J. L. Gainer and D. J. Kirwan, *Chem. Eng. Commun.*, 2003, **190**, 431.
6. N. F. Leininger, J. L. Gainer and D. J. Kirwan, *AIChE Journal*, 2004, **50**, 511.
7. R. Kumar, P. Chaudhary, S. Nimesh and R. Chandra, *Green Chem.*, 2006, **8**, 356.
8. H. X. Zhang, Y. H. Zhang, L. F. Liu, H. L. Xu and Y. G. Wang, *Synthesis*, 2005, 2129.
9. X. C. Wang, Z. J. Quan and Z. Zhang, *Tetrahedron*, 2007, **63**, 8227.
10. D. Q. Xu, S. P. Luo, Y. F. Wang, A. B. Xia, H. D. Yue, L. P. Wang and Z. Y. Xu, *Chem. Commun.*, 2007, 4393.
11. V. V. Namboodiri and R. S. Varma, *Green Chem.*, 2001, **3**, 146.
12. J. H. Li, W. J. Liu and Y. X. Xie, *J. Org. Chem.*, 2005, **70**, 5409.
13. V. Declerck, P. Ribiere, Y. Nedellec, H. Allouchi, J. Martinez and F. Lamaty, *Eur. J. Org. Chem.*, 2007, 201.
14. E. Colacino, L. Daich, J. Martinez and F. Lamaty, *Synlett*, 2007, 1279.
15. V. Declerck, J. Martinez and F. Lamaty, *Synlett*, 2006, 3029.
16. B. A. Roberts, G. W. V. Cave, C. L. Raston and J. L. Scott, *Green Chem.*, 2001, **3**, 280.
17. P. C. Andrews, A. C. Peatt and C. L. Raston, *Green Chem.*, 2004, **6**, 119.
18. F. M. Kerton, S. Holloway, A. Power, R. G. Soper, K. Sheridan, J. M. Lynam, A. C. Whitwood and C. E. Willans, *Can. J. Chem.*, 2008, **86**, 435.

19. T. R. van den Ancker, G. W. V. Cave and C. L. Raston, *Green Chem.*, 2006, **8**, 50.
20. S. L. Jain, S. Singhal and B. Sain, *Green Chem.*, 2007, **9**, 740.
21. C. K. Z. Andrade, S. C. S. Takada, P. A. Z. Suarez and M. B. Alves, *Synlett*, 2006, 1539.
22. A. Haimov and R. Neumann, *Chem. Commun.*, 2002, 876.
23. S. Chandrasekhar, C. Narsihmulu, S. S. Sultana and N. R. Reddy, *Chem. Commun.*, 2003, 1716.
24. H. F. Zhou, Q. H. Fan, W. J. Tang, L. J. Xu, Y. M. He, G. J. Deng, L. W. Zhao, L. Q. Gu and A. S. C. Chan, *Adv. Synth. Catal.*, 2006, **348**, 2172.
25. H. F. Zhou, Q. H. Fan, Y. Y. Huang, L. Wu, Y. M. He, W. J. Tang, L. Q. Gu and A. S. C. Chan, *J. Mol. Catal. A-Chem.*, 2007, **275**, 47.
26. L. K. Xu, K. H. Lam, J. X. Ji, J. Wu, Q. H. Fan, W. H. Lo and A. S. C. Chan, *Chem. Commun.*, 2005, 1390.
27. D. J. Heldebrant and P. G. Jessop, *J. Am. Chem. Soc.*, 2003, **125**, 5600.
28. J. Q. Wang, F. Cai, E. Wang and L. N. He, *Green Chem.*, 2007, **9**, 882.
29. T. J. Dickerson, N. N. Reed and K. D. Janda, *Chem. Rev.*, 2002, **102**, 3325.
30. B. Sauvagnat, F. Lamaty, R. Lazaro and J. Martinez, *Tetrahedron Lett.*, 2000, **41**, 6371.
31. J. S. Tian, C. X. Miao, J. Q. Wang, F. Cai, Y. Du, Y. Zhao and L. N. He, *Green Chem.*, 2007, **9**, 566.
32. X. Y. Dou, J. Q. Wang, Y. Du, E. Wang and L. N. He, *Synlett*, 2007, 3058.
33. J. Chen, S. K. Spear, J. G. Huddleston, J. D. Holbrey, R. P. Swatloski and R. D. Rogers, *Ind. Eng. Chem. Res.*, 2004, **43**, 5358.
34. H. C. Hsiao, S. M. Kao and H. S. Weng, *Ind. Eng. Chem. Res.*, 2000, **39**, 2772.
35. Z. Guo, G. C. April, M. Li, H. D. Willauer, J. G. Huddleston and R. D. Rogers, *Chem. Eng. Commun.*, 2003, **190**, 1155.
36. Z. Guo, M. Li, H. D. Willauer, J. G. Huddleston, G. C. April and R. D. Rogers, *Ind. Eng. Chem. Res.*, 2002, **41**, 2535.
37. S. Perrier, H. Gemici and S. Li, *Chem. Commun.*, 2004, 604.
38. K. Tanemura, T. Koike, S. Komatsu, S. Goto, Y. Nishida, T. Suzuki and T. Horaguchi, *Chem. Lett.*, 2007, **36**, 782.
39. M. N. Nadagouda and R. S. Varma, *Cryst. Growth Des.*, 2008, **8**, 291.
40. X. M. Ma, T. Jiang, B. X. Han, J. C. Zhang, S. D. Mao, K. L. Ding, G. M. An, Y. Xie, Y. X. Zhou and A. L. Zhu, *Catal. Commun.*, 2008, **9**, 70.
41. A. Corma, H. Garcia and A. Leyva, *Tetrahedron*, 2005, **61**, 9848.
42. K. Hariri, S. Al Akhrass, C. Delaite, P. Moireau and G. Riess, *Polym. Int.*, 2007, **56**, 1200.
43. K. Hariri, C. Delaite, P. Moireau and G. Riess, *Eur. Polym. J.*, 2007, **43**, 2207.

CHAPTER 9

Tunable and Switchable Solvent Systems

9.1 Introduction

For many chemical processes, there is no perfect solvent. A reaction might proceed via the preferred mechanism in a polar solvent, but a non-polar solvent would be advantageous in the work up of the reaction. A cosmetic or other consumer product may require a non-volatile solvent to ensure a long shelf life, but a certain level of volatility may be required in its use. These sorts of conundrums have led researchers to develop solvents with switchable properties. Although these media are relatively new within the alternative solvents field, they are likely to become increasingly important additions to the green solvent toolbox and many new discoveries will be made during the next decade. However, at this point it is hard to identify any particular general advantages and disadvantages of these solvents over the other solvent systems discussed in this book.

A significant advantage that these switchable solvents have over many other media is that they can be tailor-made for a particular process and particular properties can be turned on and off as desired. Unfortunately, this means that in most cases they will be considerably more expensive than simple alternatives such as water. Often the switch in these systems is the introduction of a gas such as carbon dioxide, and although the pressures involved are typically lower than those used for supercritical conditions, many users would still be wary about using and containing these gases. Further information on the switching mechanism for several cases is provided below.

RSC Green Chemistry Book Series
Alternative Solvents for Green Chemistry
By Francesca M. Kerton
© Francesca M. Kerton 2009
Published by the Royal Society of Chemistry, www.rsc.org

9.2 Chemical Examples

9.2.1 Gas Expanded Liquids

Gas expanded liquids (GXLs) and particularly CO_2 expanded liquids (CXLs) are relatively new and promising alternative reaction media.[1] They offer several advantages over traditional solvents, and require lower pressures and therefore less energy than supercritical carbon dioxide (scCO$_2$). However, clearly another liquid is needed when CXLs are used, and this is usually a petroleum sourced VOC. Because of the safety advantages of carbon dioxide over other compressable gases that are used in this field, *e.g.* ethane, this chapter focuses on CXLs.

CXLs have been used in a variety of roles including separations, particle and polymer processing and catalytic reaction media. They offer several advantages over conventional reaction media (Table 9.1).[1]

As the carbon dioxide dissolves in the organic liquid, the liquid expands volumetrically. However, not all liquids behave in the same way and therefore they have been divided into three general classes (Table 9.2). Class I liquids (*e.g.* water) do not dissolve carbon dioxide significantly and therefore do not expand much. Most traditional organic solvents are class II liquids (*e.g.* hexane and acetonitrile); they dissolve large quantities of carbon dioxide and therefore expand significantly (Table 9.2 and Figure 9.1). Ionic liquids, liquid polymers and crude oil are grouped as class III liquids and are intermediate between

Table 9.1 Summary of advantages of CXLs over conventional reaction media and scCO$_2$.

Process advantages	Ease of removal of the CO_2
	Enhanced solubility of reagent gases
	Fire suppression capability of CO_2
	Milder process pressures (tens of bars) compared to scCO$_2$ (typically > 100 bar)
Reaction advantages	Higher gas miscibility compared to ambient condition organic solvents
	Enhanced transport rates due to the properties of dense CO_2
	Between one and two orders of magnitude greater rates than in neat organic solvents or scCO$_2$
Environmental advantages	Substantial replacement of organic solvents with benign dense phase CO_2

Table 9.2 A comparison of different liquid classes and their expansion behaviour under CO_2 pressure at 40 °C.

Class	Solvent	P/bar	Volumetric expansion, %	Wt% CO_2	Mol% CO_2
I	H_2O	70	na	4.8	2.0
II	acetonitrile	69	387	83	82
III	PEG-400	80	25	16	63

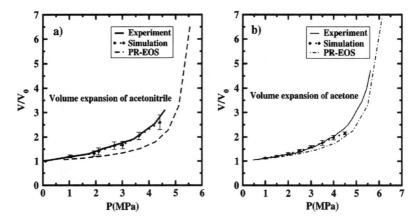

Figure 9.1 Expansions in volume of solvents: (a) acetonitrile and (b) acetone in the presence of carbon dioxide at varying pressures, 1 MPa = 1 bar. [Reprinted with permission from *J. Phys. Chem. B*, 2006, **110**, 13195–13202. Copyright 2006 American Chemical Society.]

class I and II as they expand moderately. Sample data for each class are provided in Table 9.2. Densities of CXLs vary with pressure (Figure 9.2). Further information can be found in the review article by Jessop and Subramaniam.[1]

To exploit CXLs to their full advantage, the phase behaviour of these organic–carbon dioxide mixtures needs to be understood. Experimental and theoretical studies have recently been undertaken, and reliable predictive tools are becoming available for chemists and engineers to take full advantage of the unusual properties of these solvents and exploit them.[2,3]

9.2.1.1   Solvent Properties of CXLs

For class II liquids the Kamlett–Taft π^* parameter typically drops significantly as the carbon dioxide pressure increases, whereas for class III liquids there is little change in this property with changing carbon dioxide pressure. For other solvent properties of these two classes, such as Kamlett–Taft α and β parameters, there is little change with changes in carbon dioxide pressure. However, dramatic changes are seen for the melting points of organic solids in the presence of compressed gases and this may therefore affect their behaviour in CXLs. For example, tetra-*n*-butylammonium tetrafluoroborate melts at 36 °C under 150 bar of carbon dioxide compared with 156 °C under 1 bar.[4,5]

As has already been described in Table 9.1, transport properties are enhanced in CXLs compared with conventional solvents. For example, diffusivities of solutes are enhanced up to 7-fold in carbon dioxide expanded methanol, with little effect being seen on the nature of the solute (benzene *vs* pyrazine). Therefore, it is thought that physical rather than chemical interactions are causing this phenomenon, including reduced viscosity and surface tension upon carbon dioxide addition. The solubility of solids, liquids and gases in CXLs will

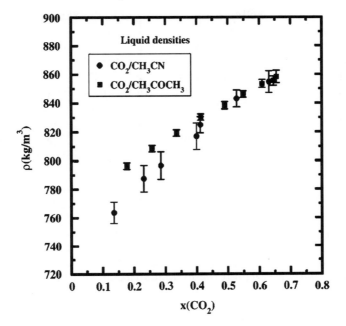

Figure 9.2 Densities of gas expanded acetonitrile (●) and acetone (■), with varying CO_2 mol fraction. [Reprinted with permission from *J. Phys. Chem. B*, 2006, **110**, 13195–13202. Copyright 2006 American Chemical Society.]

also affect chemical processes. In general, the compressed carbon dioxide in CXLs acts as an anti-solvent and can be used to induce crystallization of solutes. It can also cause pairs of miscible liquids to become immiscible upon expansion and therefore potentially separable. In contrast, the solubility of gases such as hydrogen or carbon monoxide is usually increased in CXLs compared with the unexpanded liquid. For example, the solubility of hydrogen is enhanced in ionic liquid media upon the addition of dense phase carbon dioxide, and this can lead to improved reaction rates in hydrogenation reactions in this medium.[6] However, these are generalized rules for solubility and miscibility in CXLs and there will be exceptions. One such exception is the use of carbon dioxide to trigger the mixing of two immiscible liquids. This has been observed for fluorophobic organic solvents (*e.g.* THF, cyclohexane, toluene) and fluorous solvents (*e.g.* perfluorocyclohexane).[7]

9.2.1.2 Applications of CXLs

Enhanced oil recovery (EOR) using carbon dioxide expansion is the largest scale application of gas expanded liquids.[1] EOR using carbon dioxide aids in the flushing out of oil reservoirs: carbon dioxide is injected into the well and displaces the remaining oil. It has several advantages over water, which can also be used in this process. For example, it lowers the viscosity of the crude oil, it permeates rock pores well and can be left inside the reservoir as part of a greenhouse gas mitigation strategy.

Particle formation is increasingly important in a range of areas from food-stuffs and pharmaceuticals to pigments and electronics. A large number of methods have been developed that make use of CXLs in the preparation of monodisperse fine particles (Figure 9.3).[1] Most of these techniques are descri-bed using acronyms: particles from gas-saturated solution (PGSS), gas anti-solvent (GAS), precipitation with compressed antisolvent (PCA), aerosol solvent extraction system (ASES), solution enhanced dispersion by super-critical fluids (SEDS), depressurization of an expanded liquid organic solution (DELOS) and precipitation of particles from reverse emulsions (PPRE). In many cases, adjusting the pressure of the carbon dioxide or allowing rapid expansion and loss of solvent induces the precipitation of particles. Of course, the fact that particles can be prepared in CXLs means that researchers are also investigating the processing of particles in these media, *e.g.* addition of coatings to particles. CXLs have also been used in polymer processing to make particles, and also in adjusting morphology, impregnation and mixing. In some areas, where larger particle sizes are desired, the carbon dioxide pressure can be released more slowly and recrystallizations from CXLs can be achieved. The GAS process has recently been applied to the extraction and separation of

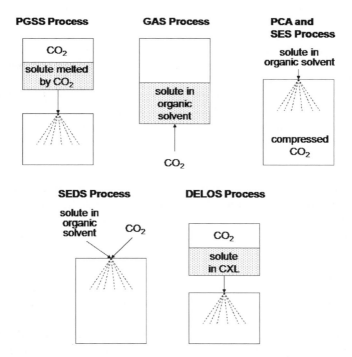

Figure 9.3   Preparation of particles using gas expanded liquids.

biorefinery chemicals including vanillin and syringol from lignin using carbon dioxide expanded methanol.[8]

Because of the better transport properties of CXLs, mixtures of carbon dioxide and organic solvents have been used as HPLC mobile phases for a range of separations (Figure 9.4). However, it is in the realm of post-reaction separations that CXLs show the most promise. Post-reaction catalyst separation usually involves changing the polarity of the liquid phase reaction medium; carbon dioxide addition offers a relatively easy way to do this. This approach has been used successfully to separate cobalt oxidation catalysts.[9] Carbon dioxide has also been used as a switch for post-reaction separations in fluorous biphasic systems, including hydrogenation and olefin epoxidations.[7,10] Further details on reactions and separations in CXLs can be found in Jessop and Subramaniam's review article.[1] Homogeneously catalysed reactions studied to date include hydrogenations, hydroformylations, oxidations and polymerizations. Reactions involving heterogeneous catalysts include hydrogenations, hydroformylations and acid catalysis. Of particular interest to green chemists are acid catalysed reactions performed in these solvents where the acid catalyst is generated *in situ* from the reaction of carbon dioxide with an alcohol or water. This has been used in catalytic acetal formation (Figure 9.5). Upon depressurization, the acid decomposes and therefore there is no acid to dispose of afterwards.[11]

In order for CXL based processes to be taken up industrially, it needs to be shown that they are economically viable or preferable to current technologies. This has been undertaken for alkene hydroformylation and compared with the current Exxon process.[12] In the CXL process, the carbon dioxide, unreacted carbon monoxide–hydrogen and olefins are separated and recycled. A polymer-bound rhodium phosphite catalyst is precipitated by adding methanol and then filtered. Cost estimations indicate that despite the expensive rhodium catalyst used in the CXL process (compared with a cheaper cobalt catalyst in the Exxon process), aldehyde production costs are comparable for the two systems. This study provides strong initial support for ongoing research in the field of CXLs and catalysis, although further ongoing collaboration between chemists and engineers is necessary.

9.2.2 Solvents of Switchable Polarity

In 2005, carbon dioxide was used to reversibly form an ionic liquid from an alcohol and the organic base 1,8,-diazabicyclo-[5.4.0]-undec-7-ene (DBU) (Figure 9.6).[13] The ionic liquid could be switched back to its neutral components by bubbling nitrogen or argon through the mixture. Switchable polarity solvents such as this have the potential to remove the requirement for changing solvents after each step of a reaction and therefore could significantly reduce the amount of solvent needed in a chemical process. In this case, the carbonate based ionic liquid that forms is polar (similar in polarity to DMF) and viscous, compared with a less viscous and less polar alcohol–base mixture (similar in polarity to chloroform). This difference in polarity can be seen when looking at

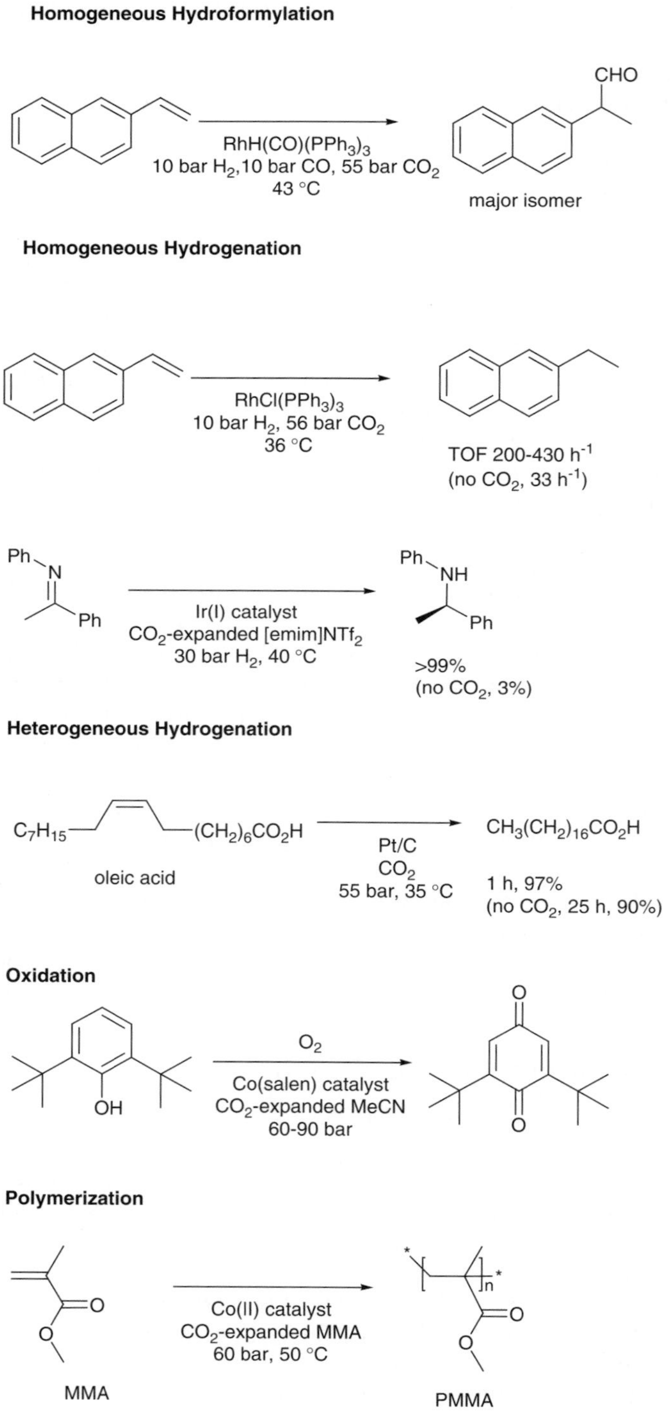

Homogeneous Hydroformylation
CHO
RhH(CO)(PPh3)3
10 bar H2,10 bar CO, 55 bar CO2
43 °C
major isomer
Homogeneous Hydrogenation
RhCl(PPh3)3
10 bar H2, 56 bar CO2
36 °C
TOF 200-430 h-1
(no CO2, 33 h-1)
Ph
N
Ph
Ir(I) catalyst
CO2-expanded [emim]NTf2
30 bar H2, 40 °C
Ph
NH
Ph
>99%
(no CO2, 3%)
Heterogeneous Hydrogenation
C7H15
(CH2)6CO2H
oleic acid
Pt/C
CO2
55 bar, 35 °C
CH3(CH2)16CO2H
1 h, 97%
(no CO2, 25 h, 90%)
Oxidation
OH
O2
Co(salen) catalyst
CO2-expanded MeCN
60-90 bar
O
O
Polymerization
O
O
MMA
Co(II) catalyst
CO2-expanded MMA
60 bar, 50 °C
n
O
O
PMMA

Figure 9.4 Some reactions performed in CXLs.

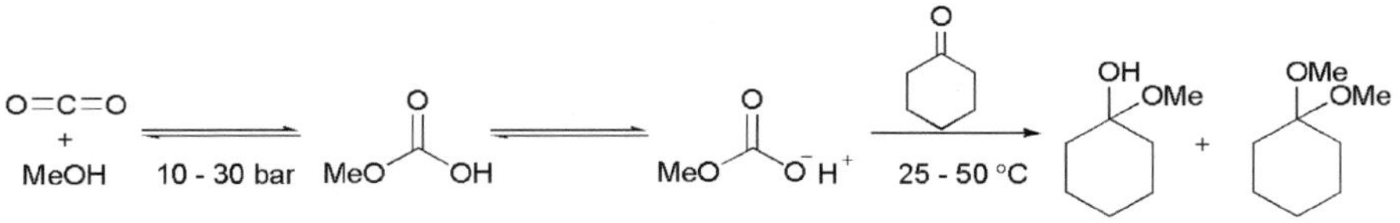

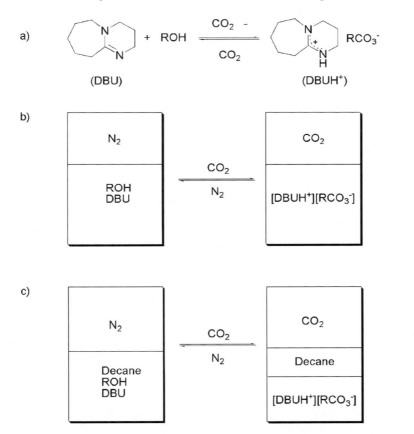

Figure 9.5 Acid catalysed acetal formation in carbon dioxide expanded methanol.

Figure 9.6 The 'switching' of a switchable solvent: (a) Reversible protonation of 1,8-diazabicyclo-[5.4.0]-undec-7-ene (DBU) in the presence of an alcohol and carbon dioxide. (b) Polarity switching in reaction (a). (c) Miscibility of decane with the alcohol–DBU mixture (non-polar) under nitrogen and separation of decane from the 'ionic liquid' (polar) under carbon dioxide. [Reprinted with permission from *Nature* 2005, *436*, 1102. Copyright 2005 Nature Publishing Group.]

the miscibility of decane with the mixture: under nitrogen it is miscible whereas under carbon dioxide it forms a separate phase. The choice of alcohol in designing this system is crucial as the corresponding hydrogencarbonate or methylcarbonate salts are not liquids, and therefore would not be suitable as

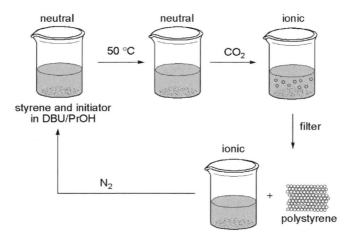

Figure 9.7 Polymerization of styrene in switchable polarity solvent consisting of DBU and 1-propanol. [Reprinted with permission from *Ind. Eng. Chem. Res.* 2008, **47**, 539–545. Copyright 2008 American Chemical Society.]

solvent systems. The most significant advantage of this system over previously mentioned tunable solvents is that the trigger to form the polar ionic liquid is just 1 atm of carbon dioxide and the less polar mixture can be slowly reformed by bubbling nitrogen through the liquid at room temperature.

Switchable polarity solvents have recently been used in two chemical syntheses:[14,15] the polymerization of styrene (Figure 9.7) and carbon dioxide–epoxide copolymerization (Figure 9.8). In the first study,[15] a non-stoichiometric mixture of DBU and 1-propanol was used as this gave a less viscous reaction mixture, which facilitated the filtration step and the isolation of the poly-styrene. The recovered ionic liquid was turned back into the less polar mixture using nitrogen and could be reused four times with the addition of some make-up solvent to replace that lost in the filtration step.

In the second study,[14] there was a desire to find a one-component switchable polarity solvent, rather than the previously studied base–alcohol mixtures. Four suitable liquid dialkylamines were discovered, which could be converted by carbon dioxide into liquid carbamates and used as switchable polarity solvents (SPS). These included dipropylamine and benzylmethylamine. Several other lighter secondary amines also afforded liquid carbamates, but these were deemed unsuitable due to their volatility and low temperature flash points. This highlights the need for finding safe alternative solvents, not just chemicals that meet the primary requirements of the investigation. Also of note is that the carbon dioxide treatment of these single-component SPS systems does not yield a single product and that the more polar form of the solvent is a mixture of carbamate ionic liquid, carbamic acid and amine. In fact, if a single product had formed, it might have been a solid and unsuitable as a solvent. The largest difference in polarity was seen for *N*-ethyl-*N*-butylamine and its polar form (after treatment with carbon dioxide); therefore, this SPS was studied in more

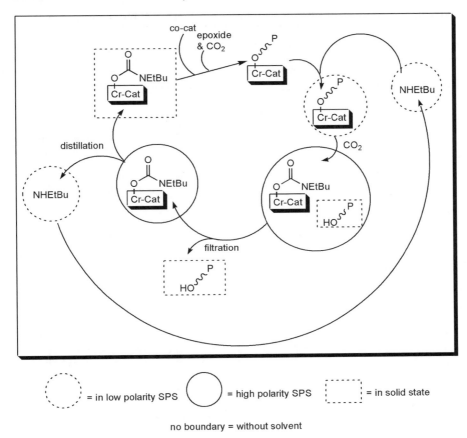

co-cat
epoxide
& CO₂

distillation

filtration

- - - - = in low polarity SPS   = high polarity SPS   [- - - -] = in solid state

no boundary = without solvent

Figure 9.8 Copolymerization of carbon dioxide in a switchable polarity solvent. [Reprinted with permission from *J. Org. Chem.* 2008, **73**, 127–132. Copyright 2008 American Chemical Society.]

detail. The solubility of various solids and liquids was studied in both forms of the SPS (Table 9.3). It was then used in the post-reaction separation of the polycarbonate formed in the catalytic copolymerization of carbon dioxide and cyclohexene oxide (Figure 9.8).[14] Both the polymer and the catalyst are soluble in the amine, but upon exposure to carbon dioxide the solvent switches to its polar form and the polymer precipitates. The chromium-containing catalyst can then be recycled and is still active in the polymerization reaction.

9.2.3 Switchable Surfactants

Jessop and co-workers used the same technology as developed for switchable solvents to obtain switchable surfactants (Scheme 9.1).[16] These have many possible applications, and their use would reduce waste and solvent usage (Table 9.4).

Table 9.3 Summary of solubility studies using *N*-ethyl-*N*-butylamine based SPS.

Soluble in both forms of SPS	Soluble in ionic form of SPS	Soluble in NHEtBu	Insoluble in both NHEtBu and ionic form of SPS
Benzylbenzamide	Tetraethyl-ammonium *p*-toluenesulfonate	Tetracosane	Cellulose
Ibuprofen	Sodium *p*-toluenesulfonate	Stilbene	Benzyl-triethylammonium chloride
Toluene			(Vinylbenzyl)-trimethylammo-nium chloride
Styrene			
Decane			
Water			

Scheme 9.1 Switchable surfactant formation from long-chain alkyl amidine, carbon dioxide and water.

Table 9.4 Potential applications of temporary or switchable surfactants.

Emulsion polymerizations
Preparation of nanoparticles
Cleaning and degreasing of equipment/metals
Enhanced oil recovery (EOR)
Oil:oil-sand separation
Viscous oil transportation
Cosmetic emulsions

Amidines mixed with water or an alcohol when exposed to 1 atm of carbon dioxide react exothermically to form a bicarbonate or alkylcarbonate salt; if the amidine bears a long chain alkyl group the resulting salt can act as a surfactant. The reaction is reversible and therefore exposure of the surfactant to argon causes the salt to release gaseous carbon dioxide and the neutral amidine reforms. This was most clearly seen by conductivity measurements (Figure 9.9) and also in photographs of emulsified hexadecane in water when exposed to carbon dioxide and separated organic–aqueous layers upon treatment with argon (Figure 9.10).

Crude oil and water mixtures also formed stable emulsions when treated with the switchable amidine carbonate surfactant, and the emulsion could be broken by exposure to argon to give two separate layers. This shows the great potential

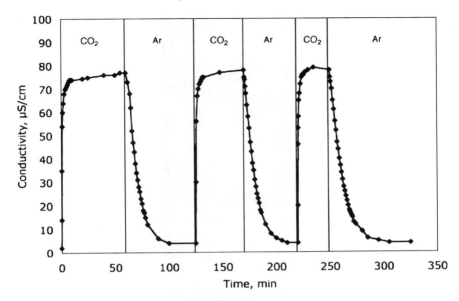

Figure 9.9   Conductivity measurements to demonstrate switching on and off of surfactant upon exposure to carbon dioxide and argon. [Reprinted with permission from *Science* 2006, **313**, 958–960. Copyright 2006 American Association for the Advancement of Science.]

such a system has in petroleum industry and equipment cleaning applications. Their application in suspension/emulsion polymerizations was also tested. A latex suspension of polystyrene particles could be prepared using the surfactant in a free radical initiated styrene emulsion polymerization in water. Treating the mixture with argon broke the suspension, which caused the polystyrene particles to precipitate.

9.2.4   Solvents of Switchable Volatility

As described in earlier chapters, the volatility of solvents is of crucial important in their applications. In 2007, the groups of Liotta, Eckert and Jessop first reported on the formation and use of piperylene sulfone (PS) as a recyclable alternative to dimethyl sulfoxide (DMSO) (Figure 9.11).[17] DMSO, in addition to dimethylformamide (DMF) and hexamethylphosphoramide (HMPA), is a widely used dipolar, aprotic solvent. However, it is difficult to remove from products by distillation and is rarely recycled. In contrast, PS decomposes cleanly at temperatures above 100 °C to give *trans*-1,3-pentadiene and sulfur dioxide, which reform PS at room temperature (Figure 9.11).

PS possesses very similar solvent properties to DMSO. For example, its $E_T(30)$ value of 189 kJ mol^{-1} is exactly the same as that of DMSO and its dielectric constant of 42.6 is of the same order of magnitude to that of DMSO, which has a dielectric constant of 46.7. In some anionic nucleophilic

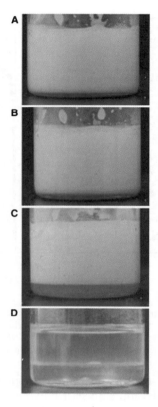

Figure 9.10   Emulsion switching for a hexadecane–water 2:1 (v/v) mixture containing switchable surfactant, after carbon dioxide treatment and 10 min shaking and (A) 5 min wait period, (B) 30 min wait period and (C) 24 h wait period. (D) After subsequent treatment with argon to 'turn off' emulsification. [Reprinted with permission from *Science* 2006, **313**, 958–960. Copyright 2006 American Association for the Advancement of Science.]

substitution reactions (Scheme 9.2), the rate of reaction in PS is of the same order of magnitude as in DMSO. However, in the case of some nucleophiles, the reactions are significantly slower in PS. It has been proposed that this slower rate is due to better solvation of the cation in DMSO, leading to greater ion pair separation, which enhances the nucleophilicity of the anion.

PS has also been used in the copper catalysed aerobic oxidation of primary alcohols (Scheme 9.3).[18] The selective oxidation of primary alcohols into aldehydes can be complicated by overoxidation to carboxylic acids or even decomposition products. These side reactions were not observed in PS, and a high turnover frequency ($>31\,h^{-1}$) was achieved. The product could be easily isolated by extraction into *n*-pentane and the PS catalyst-containing phase could be recycled three times.

It is worth noting that the volatility switch of this solvent has yet to be used to full effect in chemical reactions, and PS has primarily been used as a direct

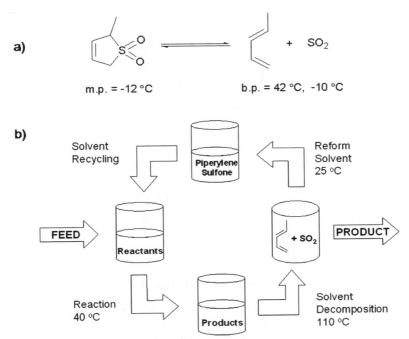

Figure 9.11 How piperylone sulfone can be used as a labile, recyclable alternative to DMSO: (a) the thermally reversible reaction to form the solvent; (b) the recycling process. [Reprinted with permission from *Chem. Commun.* 2007, 1427–1429. Copyright 2007 The Royal Society of Chemistry.]

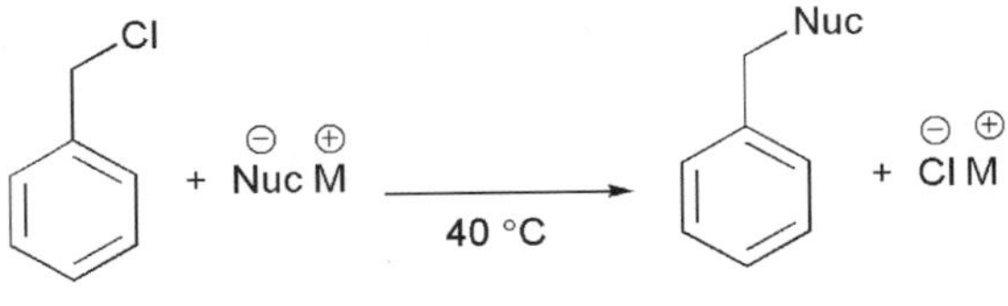

Nuc M = Potassium thioacetate, DMSO = PS: $k > 1800 \times 10^4$ l mol^{-1} s^{-1}
= Caesium acetate, DMSO: $k = 22.7 \times 10^4$ l mol^{-1} s^{-1}, PS: $k = 0.35 \times 10^4$ l mol^{-1} s^{-1}

Scheme 9.2 Comparison of nucleophilic displacement reactions in DMSO and PS.

replacement for DMSO. Therefore, it would not be surprising to see many more publications in the near future, given the recent advances in this field of switchable solvents.

9.2.5 Thermomorphic and Related Biphasic Catalysis

Solubility switching behaviour is one of the main benefits of fluorous biphasic catalysis (discussed in Chapter 7); however, other specially designed catalysts

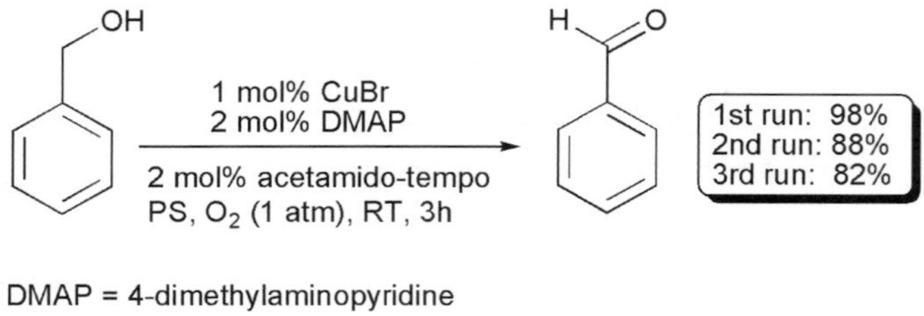

DMAP = 4-dimethylaminopyridine

Scheme 9.3 Recycling and reuse of copper catalyst and solvent in the aerobic oxidation of benzyl alcohol.

also use solubility switching to enable their facile separations, usually through changes in temperature.[19,20] A detailed discussion of these catalysts is beyond the scope of this book, but is mentioned briefly here in order to provide a full picture of the catalyst recovery field.

9.3 Summary and Outlook for the Future

Clearly this is the least mature field within the solvent alternatives arena; however, this also means that, as with tailor-made ionic liquids, it is likely that tailor-made switchable solvent systems will continue to advance and become an increasingly important area of research during the coming decades. As with all areas of clean technology, synergies and overlaps with other areas of sustainable development will increase and lead to new advances. For example, in the area of gas expanded liquids, the focus has so far been on petroleum-sourced VOCs and therefore significant advances could be made by investigating other types of gas expanded media, whether they be renewably sourced VOCs or non-volatile alternatives.

References

1. P. G. Jessop and B. Subramaniam, *Chem. Rev.*, 2007, **107**, 2666.
2. Y. Houndonougbo, H. Jin, B. Rajagopalan, K. Wong, K. Kuczera, B. Subramaniam and B. Laird, *J. Phys. Chem. B*, 2006, **110**, 13195.
3. J. L. Gohres, C. L. Kitchens, J. P. Hallett, A. V. Popov, R. Hernandez, C. L. Liotta and C. A. Eckert, *J. Phys. Chem. B*, 2008, **112**, 4666.
4. A. M. Scurto and W. Leitner, *Chem. Commun.*, 2006, 3681.
5. A. M. Scurto, E. Newton, R. R. Weikel, L. Draucker, J. Hallett, C. L. Liotta, W. Leitner and C. A. Eckert, *Ind. Eng. Chem. Res.*, 2008, **47**, 493.
6. M. Solinas, A. Pfaltz, P. G. Cozzi and W. Leitner, *J. Am. Chem. Soc.*, 2004, **126**, 16142.
7. K. N. West, J. P. Hallett, R. S. Jones, D. Bush, C. L. Liotta and C. A. Eckert, *Ind. Eng. Chem. Res.*, 2004, **43**, 4827.
8. C. Eckert, C. Liotta, A. Ragauskas, J. Hallett, C. Kitchens, E. Hill and L. Draucker, *Green Chem.*, 2007, **9**, 545.

9. M. Wei, G. T. Musie, D. H. Busch and B. Subramaniam, *J. Am. Chem. Soc.*, 2002, **124**, 2513.
10. C. D. Ablan, J. P. Hallett, K. N. West, R. S. Jones, C. A. Eckert, C. L. Liotta and P. G. Jessop, *Chem. Commun.*, 2003, 2972.
11. X. F. Xie, C. L. Liotta and C. A. Eckert, *Ind. Eng. Chem. Res.*, 2004, **43**, 2605.
12. J. Fang, H. Jin, T. Ruddy, K. Pennybaker, D. Fahey and B. Subramaniam, *Ind. Eng. Chem. Res.*, 2007, **46**, 8687.
13. P. G. Jessop, D. J. Heldebrant, X. W. Li, C. A. Eckert and C. L. Liotta, *Nature*, 2005, **436**, 1102.
14. L. Phan, J. R. Andreatta, L. K. Horvey, C. F. Edie, A. L. Luco, A. Mirchandani, D. J. Darensbourg and P. G. Jessop, *J. Org. Chem.*, 2008, **73**, 127.
15. L. Phan, D. Chiu, D. J. Heldebrant, H. Huttenhower, E. John, X. W. Li, P. Pollet, R. Y. Wang, C. A. Eckert, C. L. Liotta and P. G. Jessop, *Ind. Eng. Chem. Res.*, 2008, **47**, 539.
16. Y. X. Liu, P. G. Jessop, M. Cunningham, C. A. Eckert and C. L. Liotta, *Science*, 2006, **313**, 958.
17. D. Vinci, M. Donaldson, J. P. Hallett, E. A. John, P. Pollet, C. A. Thomas, J. D. Grilly, P. G. Jessop, C. L. Liotta and C. A. Eckert, *Chem. Commun.*, 2007, 1427.
18. N. Jiang, D. Vinci, C. L. Liotta, C. A. Eckert and A. J. Ragauskas, *Ind. Eng. Chem. Res.*, 2008, **47**, 627.
19. D. E. Bergbreiter and S. D. Sung, *Adv. Synth. Catal.*, 2006, **348**, 1352.
20. A. Behr, G. Henze and R. Schomacker, *Adv. Synth. Catal.*, 2006, **348**, 1485.

CHAPTER 10
Industrial Case Studies

10.1 Introduction

Just as the number of chemical processes that are developed using greener solvents is ever increasing, the number of industrial examples also steadily increases. This can be seen in all areas from paints, coatings and consumer products, through bulk chemical production, fine chemical synthesis and pharmaceuticals, to the extraction and processing of feedstocks in petrol and biorefineries. Green advances related to solvent use (reduction or employment of alternatives) are frequently lauded in the annual U.S. EPA Presidential Green Chemistry Challenge.[1] A particularly outstanding example was the 1997 award for a solvent free imaging system (DryView™), which at that time had already eliminated the annual disposal of 0.7 million L of developer, 1.3 million L of fixer and 200 million L of contaminated water! In the pharmaceutical industry, the idea of critically assessing their processes has risen in importance.[2] Safety and environmental issues are key reasons for this, and as was seen in Chapter 1, sensible solvent choice is often crucial in reducing risks in these areas. Additionally, in optimizing a process, an inherently better solvent is often an essential component. For example, many pharmaceutical intermediates are now isolated through enzymatic kinetic resolution and therefore the solvent that is used for at least one step in the process is water.

A wide range of different reactor types (*e.g.* continuous, membrane, bubble) have been used to perform large scale processes using alternative solvents.[3] Conventional batch reactors and extraction vessels have been used in many cases. However, process intensification is moving forward hand in hand with alternative solvents and therefore engineering solutions often have an important role to play in this field. Nevertheless, these are not discussed at length in this chapter and will probably be the subject of another book within the green chemistry series.

RSC Green Chemistry Book Series
Alternative Solvents for Green Chemistry
By Francesca M. Kerton
© Francesca M. Kerton 2009
Published by the Royal Society of Chemistry, www.rsc.org

10.2   Selected Applications: Examples

Many important large scale industrial processes are run solvent free, *e.g.* free radical polymerization of ethylene and the Haber process for the synthesis of ammonia. The latter process uses a heterogeneous catalyst and many processes that employ a heterogeneous catalyst use either gas phase or liquid phase neat reagents. Therefore, the development of new heterogeneous catalysts for industrial processes continues to be of utmost importance for a greener chemical industry. An interesting example that is currently being tackled is the manufacture of ε-caprolactam (CPL), which is a precursor to nylon-6. The industrial manufacture of this compound uses hazardous oleum (sulfuric acid) and produces 1.5 molar equivalents of ammonium sulfate per mole of CPL; it is therefore not very atom efficient. A new process has been developed that can be performed in a single step from the same precursor (cyclohexanone).[4,5] No aggressive reagents or solvents were used and the only components were the starting material cyclohexanone, air as the oxidant, ammonia as a reagent and a bifunctional nanoporous metal doped aluminophosphate (Figure 10.1). The whole reaction was performed at 80 °C and under 35 bar air pressure. The product selectivity at this time was a moderate 78% and further optimization is needed to take the reaction to the next level and make it commercially viable. However, this reaction clearly demonstrates a useful strategy for reducing solvent use: (1) use a heterogeneous catalyst and liquid or gaseous reagents and (2) use a bifunctional or even a switchable catalyst to achieve multiple transformations in a single step or pot.

At this time, I am not aware of any industrial scale solid-state synthetic procedures. However, as described in Chapter 2, kilogram scale reactions of this type have been performed,[6] and with considerable advances being made in the use of ball mills for such reactions on a laboratory scale,[7] it is only a matter of time before these innovations reach commercialization.

As outlined in Chapter 5, solvents from renewable feedstocks can be used as direct replacements for many petroleum based solvents. Therefore, they are already making a significant impact in the field of cleaning and degreasing where a lipophilic solvent is usually essential. DuPont has proposed dibasic

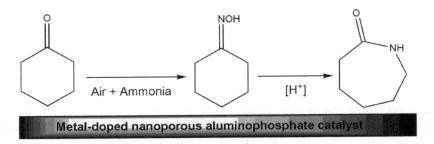

Figure 10.1   One-step solvent free catalytic production of ε-caprolactam (precursor of nylon-6).

esters (methyl esters of adipic, glutaric and succinic acids) as green solvents because of their low toxicity, carcinogenicity and volatility. They are an unwanted by-product of nylon manufacture, and are a green alternative to CH_2Cl_2 in paint strippers.[8] Renewable solvents based on methyl esters and terpenes have been used in the development of greener flexographic printing.[1] This technique is widely used in printing food wrappers and boxes and traditionally uses millions of litres of VOC solvents (*e.g.* xylene) each year.

Because of the expense of fluorous solvents, no industrial scale reactions have been reported so far. However, significant developments have been made including the use of flow reactors,[9,10] and supported-fluorous systems,[11] as described in Chapter 7. Therefore, with the continued level of interest in these reaction media, it is likely that at some point larger scale reactions will be pursued.

The rest of this chapter describes some industrial processes that use water, carbon dioxide or ionic liquids as solvents. In some cases, such as supercritical water oxidation and catalytic ionic liquids, the solvent is also a reagent.

10.2.1   Water as a Solvent and Reaction Medium

As water is abundant and non-toxic, it is an important solvent for industrial biphasic processes. Phase transfer catalysis (PTC) is well established on an industrial scale as it allows cheap inorganic bases (*e.g.* potassium hydroxide) to be used in place of organic amines. Water is used as the medium for emulsion polymerizations to produce around 10 million tonnes of polymer annually.[12] It is also a major component in the formulations of consumer goods including personal care products. Applications in the area of water based formulations and coatings have won green chemistry awards on several occasions. For example, in 2000 Bayer won an award for effectively replacing VOCs in their polyurethane coatings with water and in 1999 Nalco won an award for developing a new method for the water based synthesis of acrylamide based polymers without VOCs and surfactants.[1]

One of the best-known applications of green chemistry on a large scale is the Ruhrchemie–Rhône-Poulenc process. It possesses outstanding efficiency and produces very little waste. The synthesis of aldehydes via hydroformylation of alkenes is an industrially important process and is used to produce around 6 million tonnes of aldehydes a year.[3] Most of this is done using organic solvents. However, in 1975 a water soluble rhodium phosphine complex was discovered that could also perform this reaction and ultimately, this led to industrial scaling up as the Ruhrchemie–Rhône-Poulenc hydroformylation process. Initially, the continuous hydroformylation of propene was performed on a scale of 120 000 tonnes per year but is now at a level of 800 000 tonnes per year.[3,13,14] The process uses only gaseous substrates: propene, hydrogen and carbon monoxide. These dissolve in the aqueous phase but the product forms a separate organic phase that can be separated easily and is virtually free from rhodium contamination. The process achieves high yields and selectivities (99% butanals,

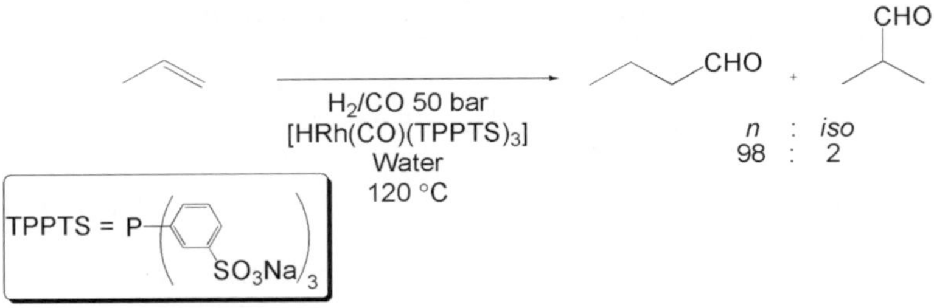

Figure 10.2 Hydroformylation of propene to *n*-butanal.

Table 10.1 Summary of environmental benefits of the aqueous biphasic Ruhrchemie-Rhône-Poulenc hydroformylation process.[3]

Use of water in place of toxic solvents
Close to zero emissions
Mild reaction conditions that lead to significant energy conservation
High selectivity towards desired linear aldehyde isomer
Very low loss of precious metal catalyst

n:iso 98:2, C_4 products >99.5%) under relatively mild conditions (120 °C, 50 bar) (Figure 10.2).[14] Since the development of this process, other types of hydrophilic phosphines have been employed for the reaction on a laboratory scale and these give higher activities and sometime better *n:iso* ratios. However, they are generally more complex structures and more expensive than TPPTS and therefore the original ligand is still used.

This process has been closely scrutinized over the last 20 years and its environmental benefits are summarized in Table 10.1.

The success of this process led to the development of other aqueous biphasic metal catalysed reactions on an industrial scale (Table 10.2), including various carbon–carbon coupling reactions (Figure 10.3). These aqueous phase organometallic reactions have been extensively reviewed.[13,14]

Biocatalysts are most often employed in aqueous solution and offer the chemist exquisite selectivity. It is therefore not surprising that they are now being employed at the industrial level, and of course water is the solvent. The application of biocatalysts to industrial chemical synthesis was recently reviewed,[15] and two examples will be highlighted here. However, enzymes in water have found use in all sectors of the chemical manufacturing industry from pharmaceuticals through fine chemicals and materials to bulk chemical production.

Talampanol (LY300164) is a drug used to treat epilepsy and neurodegenerative diseases. In an optimized procedure, the first step in its production is enzymatic in nature (Figure 10.4). *Zygosaccharomyces rouxii* is used to perform a biocatalytic reduction and excellent yields and enantioselectivities have been

Table 10.2   Commercial aqueous biphasic catalytic processes.[14]

Process	Catalyst	Products	Capacity, tonnes y^{-1}
Ruhrchemie-Rhône-Poulenc (now Celanese)	Rh-TPPTS	n-Butanal	800 000
Kururay Co. Ltd	Pd-TPPMS[a]	n-Octanol and nonadiol	5000
Clariant AG	Pd-TPPTS	Substituted biphenyls	<1000
Rhodia (formerly Rhône-Poulenc)	Rh-TPPTS	Vitamin precursors	unknown

[a]TPPMS = monosulfonated triphenylphosphine.

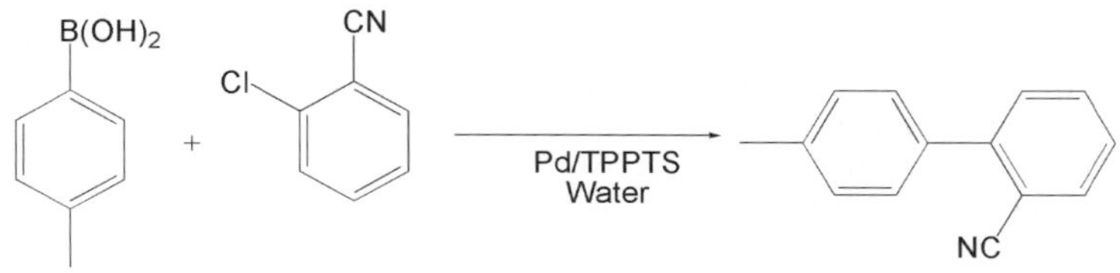

Figure 10.3   Aqueous biphasic Suzuki coupling to yield substituted biphenyls.

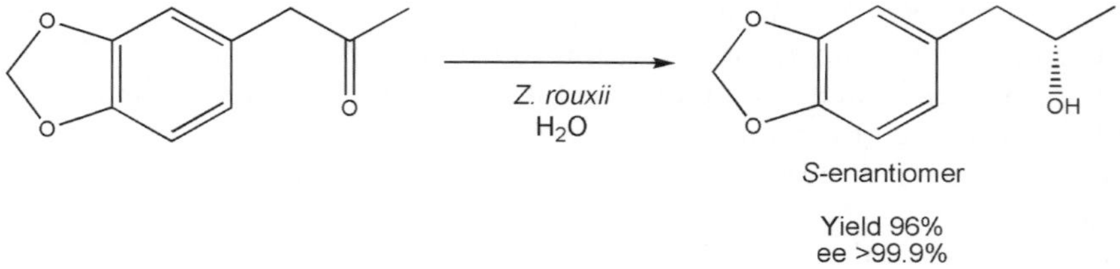

Figure 10.4   Enzymatic synthesis of key intermediate in the manufacture of the drug talampanol.

achieved. In combination with other modifications to its manufacture, this has led to a reduction in solvent use by 340 000 L per 1000 kg of product.

In the area of bulk chemical manufacturing, immobilized *Rhodococcus rhodochrous* J1 has been used to convert acrylonitrile to acrylamide. The reaction is now being performed on a scale of >40 000 tonnes per year. The yield is close to quantitative and therefore waste production is close to zero. This contrasts starkly with the traditional manufacturing route that involved hydration of acrylonitrile at 70–120 °C by Raney copper, which produced considerable amounts of toxic waste.

In addition to water under close to ambient conditions, supercritical water (SCW) is also be used on an industrial scale for SCW oxidation (SCWO)

Table 10.3 Main companies and commercial plants operating a SCWO process.[16]

Company	Commercial plant	Location	Application and capacity	Important dates
Organo Corp	Nittetsu semi-conductor factory	Japan	Waste from semi-conductors man-ufacture, $63\,kg\,h^{-1}$	Built 1998, no longer functioning
MODEC	Several companies	Germany	Pharmaceutical wastes, pulp and paper mill waste, sewage sludge, $2\,t\,day^{-1}$	Decommissioned 1996
General Atomics	U.S. Army	IN, USA	Bulk VX nerve gas agent hydrolysis, chemical agents, explosives, $949\,kg\,h^{-1}$	Commissioned 1999 (pilot plant 2000–2001)
	U.S. Army	KY, USA	Chemical agents, ageing munitions	Contract awarded 2003, expected completion 2009
Foster-Wheeler	U.S. Army	AK, USA	Chemical agents, explosives, smokes and dyes	1998, no longer functioning
Eco Waste Technologies (acquired by Chematur 1999)	Huntsman Chemical	TX, USA	Oxygenated and N-containing hydrocarbons, $1500\,kg\,h^{-1}$	1994–1999
Chematur	AquaCritox® process (pilot scale)	Sweden	N-containing wastes, $250\,kg\,h^{-1}$	1998
	Aqua Cat® Process, Johnson Matthey	UK	Platinum group metal recovery, destroy organic contaminants, $3000\,kg\,h^{-1}$	Commissioned 2002
	Municipal	Japan	Municipal sludge, $1100\,kg\,h^{-1}$	2000 (Built)
SRI International	Mitsubishi Heavy Industries	Japan	PCBs and chlori-nated wastes	2005 (Built)
Hydro-Processing	Municipal	TX, USA	Mixed waste, $9.8\,t\,day^{-1}$	2001 (Built), stopped due to corrosion

processes. A summary of the main companies and commercial plants in operation as of 2006 is provided in Table 10.3.[16] There is obviously consider-able interest in this technology from a number of sectors and innovative design of new plants is essential so that the corrosion and plugging problems that have thwarted some of the earlier plants become a thing of the past.

10.2.2 Carbon Dioxide as a Solvent and Reaction Medium

As carbon dioxide is cheap and abundant, several processes have been developed industrially that employ it as a solvent. The engineering needed for its use is well understood because of its use on a large scale in coffee decaffeination using a semi-continuous process (Figure 10.5).[3,17] The beans are pre-soaked in water to facilitate the extraction process and then enter the extraction vessel where $scCO_2$ (~ 90–$100\,°C$ and 100–250 bar) extracts the caffeine and some of the water. Beans enter at the top of the chamber and move toward the bottom over a 5 h period. To extract the caffeine continuously, the beans lower in the column are exposed to fresh $scCO_2$ that has just entered the extraction chamber; this ensures that the caffeine concentration inside the beans is always higher than in the surrounding solvent. Therefore, diffusion of the caffeine out of the bean is favoured. After the beans leave the extractor, they are dried and roasted. Recovery of the dissolved caffeine occurs in an absorption chamber. A shower of water droplets leaches the caffeine out of the $scCO_2$. The pure caffeine from this aqueous extract is then sold to soft-drink manufacturers and drug companies. The purified carbon dioxide is then recycled through the system. $ScCO_2$ is currently the preferred method for coffee decaffeination as it is

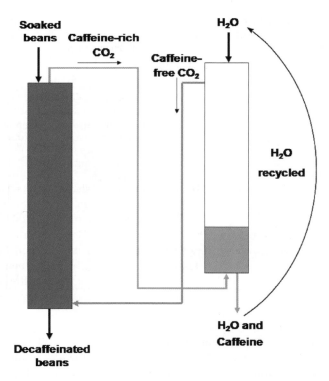

Figure 10.5   Schematic representation of the coffee decaffeination process.

Table 10.4 Industrial applications of supercritical and liquid carbon dioxide extraction.[18]

Beverages	Decaffeination (tea and coffee)
	Extracts for brewing (hops)
	Cocoa defatting
Food and flavours	Spices
	Natural colours (paprika and turmeric)
	Defatting cereal and nuts
	Vegetable oils
Cosmetics and personal care	Ginger (for toothpaste)
	Black pepper (for mouthwash)
	Paprika (for lipstick colour)
	Hempseed, wheatgerm, grape seed (for cream bases)
	Blackcurrant and borage seeds (for dietary aids)
Pharmaceutical products	
Nuclear waste	

very selective, does not leave toxic solvent residues and does not extract many of the aroma and flavour components from the beans.

As a result of the success of this process and the selectivity that carbon dioxide can enable, related extraction processes have been introduced in a number of areas (Table 10.4), one of the largest being the extraction of hop aroma for the brewing industry.[18]

As mentioned in Chapter 9, an application related to SFE is enhanced oil recovery (EOR), which uses expanded carbon dioxide on a very large scale.[19] On a smaller scale, carbon dioxide (liquid or supercritical) is being used in the dry cleaning of clothes, textile processing and metal degreasing.[20] The barriers to using this technology centre around two issues: the expense of high-pressure equipment and the poor solubility of many 'dirts' in carbon dioxide. Micell Technologies have developed equipment that uses liquid carbon dioxide (50 bar) just below ambient temperature (18–22 °C). This equipment is considerably less expensive than that needed for scCO$_2$ processes. In order to dissolve contaminants (grease, *etc.*), new and cheaper surfactants for use in carbon dioxide are continuously being developed. By 2003, over 100 000 kg of customer clothing had been cleaned in liquid carbon dioxide using Micell's equipment. Another carbon dioxide dry cleaning system, DryWash™, has been developed by Raytheon Environmental Systems and Los Alamos National Laboratory. This system uses jets of liquid carbon dioxide to agitate clothing. The use of carbon dioxide in both these methods reduces the environmental burden of dry cleaning and also the worker and consumer health issues associated with the use of perchloroethylene.

Industrially, scCO$_2$ has been used extensively in polymer processing and synthesis. During the last 10 years, DuPont built a plant that can produce 1000 tonnes of Teflon™ and other fluoropolymers per year.[21,22] The polymers produced in this plant are claimed to have superior performance and processing capabilities. Carbon dioxide is seen as the most viable industrial solvent for fluoropolymer synthesis, as hydrocarbon solvents can cause detrimental side reactions and CFCs that were in common use in this field are now prohibited in most locations and for most applications. In the area of polymer processing, a supercritical carbon dioxide fluid spray process (UNICARB process) was commercialized in 1990 by Union Carbide with the aim of reducing the concentration of VOCs in coating formulations.[20] The UNICARB spray solution consists of 10–50 wt% dissolved carbon dioxide in the coating material. The amount of carbon dioxide used in any given application depends upon the carbon dioxide solubility, the viscosity, the solids level, the pigment loading of the coating formulation, and the spray pressure and temperature. In this process, carbon dioxide acts as a good viscosity reducer and allows a novel mechanism for atomization. In this mechanism, as the dissolved carbon dioxide in the spray solution leaves the nozzle, it undergoes a rapid decompression due to the sudden pressure drop. This pressure drop creates a large driving force for nucleation and coagulation. This results in the rapid formation of small liquid droplets in the expansion zone within a short distance from the spray opening, instead of downstream from the nozzle as in a normal spray coating process. It has been shown that the UNICARB process can produce fine droplets within the same range as conventional spray systems, but with a narrower size distribution that improves the appearance of the coating. This added value that the carbon dioxide process offers over traditional spray coating is in no small part the reason why this process has been so successfully commercialized. It has been demonstrated that the UNICARB process can be used to apply a wide variety of high-quality coatings (clear, pigmented, and metallic). Union Carbide has indicated that with such developments in the coating industry, the use of organic solvents in this area can be eliminated to produce zero-VOC coatings for most applications.

Also in the area of polymer processing, sustainable polymer foaming using high-pressure carbon dioxide was recently reviewed.[23] Currently, the main production method for polymer foams is the so-called thermally induced phase separation (TIPS) process, where the foaming agent (a low boiling organic solvent such as pentane) is dissolved in the polymer and then heated. However, TIPS and other conventional foaming methods lead to materials containing harmful residual solvents. It has been estimated that in Europe in 2010, polymer foam production will lead to 256 000 tonnes of VOC emissions.[23] Therefore, significant efforts are being made to reduce the solvent demand in these processes by using either nitrogen or carbon dioxide as the foaming agent. Dow Chemical Company won a 1996 green chemistry award for their development of carbon dioxide as a blowing agent for polystyrene foam sheet packaging.[1] Their technology eliminated the use of 1.75 million tonnes of CFC or HCFC solvents per year at a full scale commercial facility. Additionally, BASF has scaled up a carbon dioxide based foaming process to produce a material called Styrodur® (an extruded polystyrene). The market for such solvent free materials is large and BASF claim that 25 million m^2 of Styrodur® is installed in Europe per year as an insulating (construction) material.[24]

Also in the area of materials for the construction industry, there are growing concerns about the durability, maintenance, production and life-cycle of cement and concrete.[25] It is thought that the principles of green chemistry could help widely in this area. For example, scCO$_2$ is being used to accelerate the natural ageing reactions of Portland cement.[20,25] This treatment alters the bulk properties of the cement, producing changes in both its structure and chemical composition through a carbonation reaction. The treated cements have enhanced physical properties including reduced porosity, permeability and pH, and increased density and compressive strength. It has also been shown that scCO$_2$ treatment allows the replacement of some Portland cement powder with inexpensive materials such as fly ashes. Although some of the materials being studied are aimed at high cost markets (such as encapsulating materials for nuclear sites), one industrial process being commercialized by Supramics involves the combination of CO$_2$ and fly ash to modify cement for low cost building materials.[20]

ScCO$_2$ has also had a significant industrial impact on the manufacture of high tech materials. The U.S. Environmental Protection Agency estimates that each day a typical chip-fabrication plant generates 15 million L of waste water and consumes thousands of litres of VOCs and corrosive chemicals. In 2002, SC Fluids Inc. was awarded a green chemistry award for their supercritical carbon dioxide resist removal (SCORR) process that uses scCO$_2$ in the processing of semiconductor wafers. The carbon dioxide is used within a closed loop and has dramatically reduced solvent, water and chemical use in semiconductor manufacture. More information on this process is available on the Green Chemistry Resource Exchange or EPA websites.[1,26]

One of the most successful commercializations of scCO$_2$ technology was undertaken by Thomas Swan & Co. Ltd in collaboration with the University of Nottingham.[27,28] The resulting continuous hydrogenation process has resulted in many prizes for the industrial–academic team. The reaction chosen for commercial exploitation was the hydrogenation of isophorone to trimethylcyclohexanone (Figure 10.6), as there was market demand for a high quality product.[28] However, the plant is multipurpose and the reactions easily can be switched by changing the catalysts within the reactor.

The Thomas Swan & Co. plant went on stream in June 2002 and produces 1000 tonnes of product per year. As can be seen in Figure 10.6, the purity of the

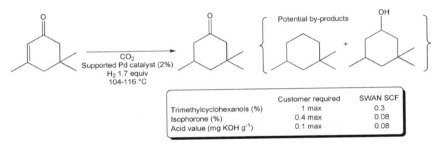

	Customer required	SWAN SCF
Trimethylcyclohexanols (%)	1 max	0.3
Isophorone (%)	0.4 max	0.08
Acid value (mg KOH g^{-1})	0.1 max	0.08

Figure 10.6   Continuous catalytic production of trimethylcyclohexanone in scCO$_2$.

product is very high, so in addition to removing the use of a VOC in the reaction, significant amounts of VOC have been removed downstream because no purifications are required. From their experiences with this process, Poliakoff, Ross and co-workers are confident that many other catalytic processes should be readily amenable for scale up in this plant.

10.2.3   RTILs in Industry

The application of ionic liquids in the chemical industry has recently been reviewed.[29] It was noted that the security of supply should not be an issue for industrial (and other) users as there are now many manufacturers able to supply RTILs on a multi-tonne scale. However, the cost of some RTILs will possibly inhibit their use on a large scale. Nevertheless, in some applications (Table 10.5) this cost will be less of an issue and in other cases less expensive choline based or alkylammonium derived salts may be an option.

Of some concern in this field is the issue of intellectual property, as there are a large number of patents protecting the area in terms of the preparation, identities and uses of these solvents. For example, one patent claims broad coverage of cheap phosphonium ionic liquids,[30] and this might prevent their use in many applications. However, the large number of patents also demonstrates significant industrial interest in these media and industrial processes using them are now online.[29]

The most successful and best known industrial process using ionic liquid technology is the biphasic acid scavenging utililizing ionic liquids (BASIL™) process of BASF, Germany.[29] It is being performed on a multi-tonne scale and demonstrates the practical handling of ionic liquids on a large scale. However, the ionic liquid is not the solvent for the reaction. In contrast, during 1996–2004 the Eastman Chemical Company operated a plant using a phosphonium ionic liquid as the solvent for the isomerization of 3,4-epoxybut-1-ene to 2,5-dihydrofuran

Table 10.5   Current and future applications of RTILs.[29]

Solvents and catalysts	Biological uses	Electrochemistry	Engineering and processing	Analytics
Synthesis	Biomass processing	Electrolytes in batteries	Coatings	Matrices for mass spectrometry
Catalysis	Drug delivery	Metal plating	Lubricants	GC columns
Microwave chemistry	Biocides	Fuel cells	Plasticizers	HPLC stationary phases
Nanochemistry	Personal care	Electro-optics	Dispersing agents	
Multiphase reactions and extractions	Embalming	Ion propulsion	Compatibilizers	

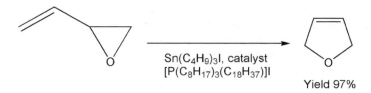

Figure 10.7 Eastman Chemical's IL based isomerization process.

Table 10.6 Summary of key advantages of Difasol (ionic liquid) process over the Dimersol (solvent free) process for alkene dimerization.[29]

Catalyst cost, use and disposal is reduced
Better dimer selectivity (>90%)
Higher and quicker yield of desired product (81 wt% conversion in 2 h vs 17 wt% in 8 h)
Potential for dimerizing higher olefins
Smaller reactor size

(Figure 10.7). This plant is currently idle because of a decrease in market demand for the product.[29]

The French Petroleum Institute has developed an ionic liquid based process for the dimerization of alkenes (Dimersol process) and it has been patented as the Difasol process. Interestingly, it can be retrofitted and operated in existing Dimersol plants. However, its biphasic nature offers several advantages over the traditional, homogeneous Dimersol process (Table 10.6).

Many petrochemical companies hold extensive patent portfolios relating to ionic liquid technologies. However, the first of these to announce an industrial process is PetroChina. The process for alkylation of isobutene uses an aluminium(III) chloride based ionic liquid and is called Ionikylation. After success at the pilot plant stage, the technology is currently being retrofitted into an existing sulfuric acid alkylation plant in China with an output of 65 000 tonnes per year. This retrofit will increase yield and capacity at the site and is the largest commercial use of ionic liquids reported to date.[29]

As discussed in Chapter 6, ionic liquids have great potential as media for electroplating and therefore researchers in the Green Chemistry Group at the University of Leicester formed a spin-off company called Scionix.[31] Scionix with Whyte Chemicals have developed a chromium electroplating process, based on choline chloride–Cr(III) derived ionic liquids,[32] that is currently operational on a pilot plant scale.

10.3 Summary and Outlook

Industrial applications using nearly all types of alternative solvents have been successful and therefore there is no reason to doubt that more successes are on the horizon. Water will remain a very popular solvent with industry, particularly for biocatalytic procedures that are often more enantioselective, and therefore more desirable, than their metal catalysed counterparts. However, as far as I am aware, simple 'on water' (non-PTC) organic transformations have not yet entered industrial use, where engineering issues such as mass-transfer may be more of a concern than in an academic laboratory. ScCO$_2$ has found a niche in industrial materials processing and continues to be an important solvent for extraction in the food and flavour industry. The results of Thomas Swan & Co. have shown that scCO$_2$ can also be employed as an effective solvent in heterogeneous catalytic processes and products of excellent quality can be obtained. Ionic liquids are increasing in large scale use, largely as a result of their commercial availability. Their key industrial applications appear to be in areas where the ionic liquid is not just a solvent, such as Lewis acid catalysed processes and electrodeposition of metal coatings.

References

1. United States Environmental Protection Agency, Presidential Green Chemistry Challenge Awards Previous Winners, http://www.epa.gov/greenchemistry/pubs/pgcc/past.html, accessed June 2008.
2. M. Butters, D. Catterick, A. Craig, A. Curzons, D. Dale, A. Gillmore, S. P. Green, I. Marziano, J. P. Sherlock and W. White, *Chem. Rev.*, 2006, **106**, 3002.
3. D. J. Adams, P. J. Dyson and S. J. Taverner, *Chemistry in Alternative Reaction Media*, John Wiley & Sons, Chichester, 2004.
4. R. Mokaya and M. Poliakoff, *Nature*, 2005, **437**, 1243.
5. J. M. Thomas and R. Raja, *Proc. Natl. Acad. Sci. U.S.A.*, 2005, **102**, 13732.
6. G. Kaupp, *CrystEngComm*, 2006, **8**, 794.
7. B. Rodriguez, A. Bruckmann, T. Rantanen and C. Bolm, *Adv. Synth. Catal.*, 2007, **349**, 2213.
8. N. E. Kobb in *Clean Solvents: Alternative Media for Chemical Reactions and Processing* (ACS Symposium Series), ed. M. A. Abraham and L. Moens, Washington, DC, 2002.
9. A. Yoshida, X. Hao, O. Yamazaki and J. Nishikido, *QSAR Comb. Sci.*, 2006, **25**, 697.
10. A. Yoshida, X. H. Hao and J. Nishikido, *Green Chem.*, 2003, **5**, 554.
11. E. G. Hope, J. Sherrington and A. M. Stuart, *Adv. Synth. Catal.*, 2006, **348**, 1635.
12. S. Mecking, A. Held and F. M. Bauers, *Angew. Chem. Int. Edit.*, 2002, **41**, 545.
13. B. Cornils and W. A. Herrmann (Ed.), *Aqueous-Phase Organometallic Catalysis*, Wiley-VCH, Weinheim, 2004.
14. E. Wiebus and B. Cornils, in *Catalyst Separation, Recovery and Recycling*, ed. D. J. Cole-Hamilton and R. P. Tooze, Springer, Amsterdam, 2006.
15. N. Q. Ran, L. S. Zhao, Z. M. Chen and J. H. Tao, *Green Chem.*, 2008, **10**, 361.

16. M. D. Bermejo and M. J. Cocero, *AIChE J.*, 2006, **52**, 3933.
17. S. N. Katz, *Scientific American*, 1997, **276** (June), 148.
18. R. Marriott, Botanix Ltd, SCF processing for UK Industry Conference, Burton-on-Trent, 2000.
19. P. G. Jessop and B. Subramaniam, *Chem. Rev.*, 2007, **107**, 2666.
20. J. M. DeSimone and W. Tumas (ed.), *Green Chemistry Using Liquid and Supercritical Carbon Dioxide*, Oxford University Press, Oxford, 2003.
21. M. McCoy, *Chem. Eng. News*, 1999, **77** (June 14), 11.
22. S. L. Wells and J. DeSimone, *Angew. Chem. Int. Ed.*, 2001, **40**, 519.
23. L. J. M. Jacobs, M. F. Kemmere and J. T. F. Keurentjes, *Green Chem.*, 2008, **10**, 731.
24. BASF, http://www2.basf.de/en/produkte/kstoffe/schaum/styrodur, accessed June 2008.
25. J. W. Phair, *Green Chem.*, 2006, **8**, 763.
26. ACS Green Chemistry Institute, http://www.greenchemex.org/, accessed June 2008.
27. R. Ciriminna, M. L. Carraro, S. Campestrini and M. Pagliaro, *Adv. Synth. Catal.*, 2008, **350**, 221.
28. P. Licence, J. Ke, M. Sokolova, S. K. Ross and M. Poliakoff, *Green Chem.*, 2003, **5**, 99.
29. N. V. Plechkova and K. R. Seddon, *Chem. Soc. Rev.*, 2008, **37**, 123.
30. A. J. Robertson, Preparation of phosphonium salts as ionic liquids, PCT Int. Appl. 0187900.
31. Scionix Ltd, http://www.scionix.co.uk/, accessed June 2008.
32. A. P. Abbott, G. Capper, D. L. Davies and R. K. Rasheed, *Chem. Eur. J.*, 2004, **10**, 3769.

Subject Index